ATOMS AND ELEMENTS

THE HISTORY
OF SCIENTIFIC IDEAS

THE ANCESTRY OF SCIENCE
(in four volumes)
The Fabric of the Heavens
The Architecture of Matter
The Discovery of Time

in preparation
Science and its Environment
A Guide to the History of Scientific Ideas

Monographs
FORESIGHT AND UNDERSTANDING
An enquiry into the aims of science

QUANTA AND REALITY
A symposium

THE BIRTH OF MATHEMATICS IN THE AGE OF PLATO

THE ANTICIPATION OF NATURE

INVESTIGATIONS INTO GENERATION

ATOMS AND ELEMENTS
A Study of Theories of Matter in England in the Nineteenth Century

ATOMS AND ELEMENTS

A Study of Theories of Matter in England in the Nineteenth Century

★

DAVID M. KNIGHT

HUTCHINSON OF LONDON

HUTCHINSON & CO (*Publishers*) LTD
178–202 Great Portland Street, London W1

London Melbourne Sydney
Auckland Bombay Toronto
Johannesburg New York

First published 1967

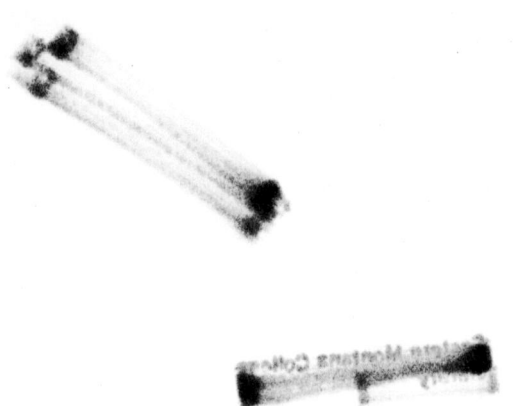

© David Knight 1967

This book has been set in Bembo, printed in Great Britain on Antique Wove paper by Anchor Press, and bound by Wm. Brendon, both of Tiptree, Essex

Acknowledgements

It is a pleasure to acknowledge the help given by a number of scholars and librarians. And, in particular, I should like to thank Dr A. C. Crombie and Mr Rom Harré of the University of Oxford; Dr W. H. Brock of the University of Leicester; and Mr Arnold Thackray of Cambridge University, for help given at various stages in the history of this work. I should also like to thank Professor Stephen Toulmin for his helpful remarks on seeing the draft of this book.

Contents

	Foreword	1
1	The Legacy of the Past	5
2	Mr Dalton and his Critics	16
3	Boscovicheans and Sceptics	37
4	Some Theories of Matter	60
5	Chemical Molecular Theories	83
6	The Debates	105
7	The End of the Affair	127
	References and Bibliography	153
	Index	163

Foreword

In the history of science, as in other branches of history, it is necessary from time to time to indulge in revision, and to see whether the general picture we have inherited from the past is a satisfactory account of what actually happened, or is essentially myth. This is perhaps particularly so in the history of science, which has often been written in a whiggish manner as an account of the *progress* of science. Authors whose views seem an anticipation of modern theories are exalted, although usually the anticipation is far from complete, and the old and new theories were designed to explain different sets of phenomena. So it has been with atomism. Since, it is argued, the world is composed of atoms, those who in Antiquity or in modern times wrote in support of atomism were, in some strong sense, right; while those who opposed them were reactionary and wrong, or at least to be apologised for. This is absurdly unhistorical. If we are to assess scientists of the past, we must judge their views not by this kind of criterion but by their consistency and their power to explain the phenomena then known and felt to be puzzling. We shall then find that some atomic theories (for to some extent every atomic theory which explains new phenomena is a new and different theory) were sound and well based; while others were naïve and speculative.

In an epoch of atomic weapons we tend to take it for granted that matter is made up, in some sense, of atoms; and to forget that the status of these entities is not quite the same as that of tables and chairs. Chemistry books and short histories of science, hurrying through the development of the subject, usually imply that we owe this understanding of the nature of matter to John Dalton who, early in the nineteenth century, revived the atomic theory of Lucretius and made it the central principle of chemistry. One would therefore expect *a priori* that the century between the publication of Dalton's theory and the discovery of atomic disintegration would have been the great classical period of atomism.

In fact this is not at all what is revealed by an inspection of the literature. Atomism was resurrected in the seventeenth century; and the nineteenth stands out as a century in which scientists were deeply divided over atomic explanations. There was no one classical, received atomic theory, but rather a number of theories overlapping in their explanatory ranges; and Dalton's was a precursor of that of Kekulé rather than of twentieth-century atomism.

The historian of atomic theories must therefore seek to keep clear a number of issues which were muddied by the protagonists themselves. There is first the question as to whether an atomic theory is wanted at all. Opponents of all atomic theories fell into two groups: those, relatively few, who believed matter to be ultimately a continuum, and those who held that the whole question was a metaphysical and not a scientific one. Science, on this view, should aim simply at equations connecting observables. Distinguished men opposed atomism on this ground until well into the first decade of the twentieth century. And until atomic theories began, towards the end of the nineteenth century, to have a 'cash value' in terms of empirical usefulness in the sciences, and to generate experimentally testable consequences, it was not unreasonable to describe them as metaphysical.

But our study is not only concerned with scepticism but also with the 'reduction' of chemistry to physics. These two sciences were, at the beginning of our period, separate and distinct. And until the 1870s it seems fair to say that chemists as a group took no notice of physical evidence; and that physicists paid scant attention to chemistry. Even among atomists, therefore, we find divisions; particularly between chemists who wanted an atomic theory to explain the laws of chemical composition, and physicists who sought to account for the physical properties of matter—its hardness, mass, elasticity, and so on. Chemists tended therefore to employ theories in which the atoms were the smallest units of the various elements, although this meant that there were many different kinds of atom; while physicists thought in terms of basic units of matter, which would compose the different chemical elements according to their different arrangements. Many chemical atomists, like Kekulé and Williamson, denied all interest in physicists' atomic theories, or theories of matter in general, and argued that chemical atomism, a theory without hypothesis, should be accepted because it worked, whatever the ultimate structure of matter turned out to be. The achievement of an atomic theory acceptable to both chemists and physicists, in which the 'atoms' of the chemical elements were interpreted as made up of protons and electrons, closed the gulf between the two sciences.

The study will begin with a survey of the atomic theories which nineteenth-century scientists inherited from their predecessors, and then follow the development of these theories and the criticism to which they were subjected throughout the century. It is clearly impossible, in what is still largely unmapped territory, and in a small compass, to be exhaustive. And from about 1870 the story will be told even more sketchily, for then the modern period of atomism seems in a sense to begin, and the men and theories are more familiar. The story will be largely confined to what happened in

England; but although there are undoubtedly national and regional schools of scientists as there are of painters, what happened in this country seems to have been not untypical. The swing against atomic explanations which reached a peak about the 1860s appears to have been general; and England was not, after all, quite cut off from Continental influences. A similar pattern will probably be found by students of atomism in other countries, although the details will of course be different.

Scientists of the early nineteenth century were not called natural philosophers for nothing, as Cohen and Jones emphasise in their anthology, *Science Before Darwin*. Among the physicists and chemists of this period we find the 'public commitment to general truth', and the search for Order, Law, Organisation, and Meaning to which these authors point. At a distance of more than a century we can often take for granted the discoveries of such scientists as Davy and Faraday, and turn our attention to their arguments and metaphysical assumptions, which are perhaps more generally interesting than the details of experimental procedures. This will be, however, an internal history of a period of scientific thought. We shall be concerned with how scientists reacted to and developed the thoughts of other scientists, rather than with more remote and intangible influences, or those economic circumstances which some are able to see as playing an important role in the development even of the theoretical sciences of the last century. Progress or change in this field came about through dialogue, frequently with predecessors; and following this example we shall not hesitate on occasion to argue with the dead.

DAVID KNIGHT

Durham, May, 1966

1

The Legacy of the Past

SCIENTIFIC theories are very rarely born out of the blue, and the atomic theories of the nineteenth century have their roots in the writings of scientists of the seventeenth and eighteenth centuries. 'The corpuscular philosophy' was the name given to the atomic theory as it was developed in the later seventeenth century, mostly by English scientists and philosophers who derived their ideas partly from Gassendi, and ultimately from the atomic theories of Antiquity, as expounded by Epicurus and in Lucretius' *de Rerum Natura*. For scientists in the nineteenth century the *locus classicus* of the corpuscular philosophy was the Queries which Newton appended to the later editions of his *Opticks*; in which he proposed in the form of questions hypotheses as to the nature of matter, light, and gravity which did not seem susceptible of proof.

Of atoms, Newton wrote, in a passage which despite its familiarity merits a long quotation:

> 'All these things being consider'd, it seems probable to me, that God in the Beginning form'd Matter in solid, massy, hard, impenetrable, moveable Particles, of such Sizes and Figures, and with such other Properties, and in such Proportion to Space, as most conduced to the End for which he form'd them; and that these primitive Particles being Solids, are incomparably harder than any porous Bodies compounded of them; even so very hard, as never to wear

or break in pieces; no ordinary Power being able to divide what God himself made one in the first Creation. . . . And therefore, that Nature may be lasting, the Changes of corporeal Things are to be placed only in the various Separations and new Associations and Motions of these permanent Particles; compound Bodies being apt to break, not in the midst of solid Particles, but where those Particles are laid together, and only touch in a few Points.'[1]

If the atoms were to wear away, the laws of nature would change. So the atoms must be like ideal billiard-balls: solid, massy, hard, impenetrable, and movable, because these qualities were the primary ones. Other qualities such as colour, smell, taste, and elasticity were secondary; that is, they were not found in all bodies, or were present in varying degrees, and were therefore judged to be produced by the aggregation of the particles in various ways.

This brings us to the important question: what is the object of an atomic theory? The answer would seem to be that such a theory must explain the complex and manifold properties of the various things that there are in the world, by showing that they may arise from different arrangements of rather few different kinds of particles—the atoms—having relatively simple properties. And changes must be explained in terms of mere rearrangements of the simple, permanent particles. In the corpuscular philosophy the particles had only the primary qualities, so this theory did, so far, produce the required simplification. But the problem remained of showing how the atoms could be arranged so as to produce the richly varied world that we see.

It had been suggested in Antiquity that atoms might be held together by hooks, but Newton would have none of that. He proposed, in place of such question-begging explanations, that

> 'Particles attract one another by some Force, which in immediate Contact is exceeding strong, at small distances performs the chymical Operations above-mention'd, and reaches not far from the Particles with any sensible Effect.'[2]

Newton declared that here, as in his *Principia*, he used the term 'attraction' in a purely descriptive way. He believed that some kind of pushing or pulling mechanism must be there behind the appearances:

> 'There are therefore Agents in Nature able to make the Particles of Bodies stick together by very strong Attractions. And it is the Business of experimental Philosophy to find them out.'[3]

Elsewhere Newton implied that the ultimate particles are all identical; not particles of gold or iron or phosphorus, but simply particles of matter. On the other hand, Leibniz rejected the atomic theory because it seemed impossible to believe that God would have created two or more atoms exactly alike, or indiscernible. The ultimate particles of the corpuscularians, arranged in various ways, formed the smallest units that could be found of the various substances investigated by the chemists. There was thus a hierarchy, with atoms at the bottom, building up the more complex particles of the simpler bodies we ordinarily meet, and then composing the compound bodies. The atoms might be believed to be all identical; or if they were thought to differ, it was only in, for example, size and shape (as the botanist Nehemiah Grew thought), and not in chemical properties.

The passages from Newton raise most of the questions which puzzled his successors in the hundred years after his death in 1727. Men differed over whether the various 'attractions', atomic, gravitational, electrical, and magnetic, were all the same, or at least manifestations of one power, or were irreducibly different. They also differed in their explanations of heat. In England, at least in the seventeenth century, heat had been supposed a kinetic phenomenon. Thus John Locke believed that heat was a very brisk agitation of the insensible parts of the object, which produced in us that sensation from which we denominated the object hot; so what in our sensation was *heat*, was in the object nothing but *motion*.[4] By 1800, this simple atomic view of heat had passed out of favour, and was regarded as old-fashioned. Its successor was the caloric theory, according to which heat was a weightless fluid which filled the gaps between atoms. Dalton thus believed that each atom was surrounded by an atmosphere of caloric, and that in gases as in solids and liquids the atmospheres of adjoining atoms were in contact; there was no void space between atoms.

Both the kinetic and caloric views of heat explained the expansion of bodies when they are heated; on the former theory, the atoms vibrate harder and push one another apart, and on the latter more caloric enters the pores of the substance and the pressure separates the atoms further. In the writings of Black and Lavoisier caloric was treated as a chemical substance which formed compounds like any other. The theory then seemed to give a more natural explanation than its kinetic competitor of the phenomena of latent heat, and of the emission of heat in chemical changes. Water absorbs great quantities of heat as it boils, on this view, because steam is a compound of water and caloric; and caloric can be given off in a reaction just as hydrogen, for example, can be.

For our purposes it is sufficient to remark that both these theories of heat presupposed the existence of atoms of some kind. The conflict between the two theories in about 1800 should also not be exaggerated; Garnett, Davy's predecessor at the Royal Institution in London, declared that the term 'caloric' was theory-free, and accorded with every opinion.[5] It is also worth noting that the kinetic theory of heat did not necessarily require a kinetic theory of gases. The corpuscles of gases, in Newton's view, were not rushing about all through the containing vessel; their movements were small-scale ones.

Newton's model seemed, like his work on gravity, to involve action at a distance across a void space; a notion repugnant to many, who believed that bodies cannot act where they are not. But this was not an essential feature of the corpuscular philosophy, which was attacked in the eighteenth century on two fronts: by chemists who felt that the problems of explanation in their science were being glossed over; and by the mathematical physicist, Boscovich, whose atoms possessed even fewer properties than those of the corpuscularians. For Boscovich the atom has degenerated to a mere mathematical point at the centre of a field of force; the force is all we can ever know.

The corpuscular philosophy gives the impression, though Boyle was the most important of its originators, of having been produced by physicists interested in such problems as hardness, density, optical behaviour, and the phenomena of heat. For these the theory was capable of yielding tolerably convincing explanations. But for the phenomena of chemistry such explanations as were advanced were no more than sketches in principle of how the problem might, at some date in the future, be accounted for in detail in terms of the theoretical models used in physics. It may be helpful to have such explanations; for instance theoretical chemists today explain chemical reactions in terms of electrons, orbitals, and wave equations even though these theories can provide only *post hoc* explanations, rather than predictions, in all but the simplest cases. But the electronic theory of valency is extraordinarily detailed by comparison with the corpuscular philosophy.

Since for chemists such as Boyle the corpuscles of all bodies were composed of probably the same atoms differently arranged, alchemy was for corpuscularians not an impossibility but simply a deeper-level chemical change. Nineteenth-century corpuscularians like Humphry Davy for the same reason rejected the notion that the chemical elements were irreducibly different. And one finds throughout the eighteenth century quite reputable believers, or half-believers, in alchemy. In 1811 Davy declared that 'to enquire whether the metals be capable of being decomposed and composed is a grand object of true philosophy',[6] to be carefully distinguished from the delusions of those who sought the 'powder of projection'.

Although Boyle's interest lay in chemistry more than in physics, he sometimes gives the impression of an airy physicist who felt that all chemical problems had in principle been solved. Thus he wrote in *The Sceptical Chymist*:

> 'To be short, as the difference of bodies may depend merely upon that of the schemes whereinto their common matter is put; so the seeds of things, the fire and other agents ... partly by altering the shape or bigness of the constituent corpuscles of a body, partly by driving away some of them, partly by blending others with them, and partly by some new manner of converting them, may give the whole portion of matter a new texture of its minute parts, and thereby make it deserve a new and distinct name. So that according as the small parts of matter recede from each other, or work upon each other, or are connected together after this or that determinate manner, a body of this or that denomination is produced, as some other body happens thereby to be altered or destroyed.'[7]

This passage simply shows that atomic explanations are possible in chemistry; Boyle could be said to be making the world safe for atomists, rather than producing atomic explanations himself.

Chemistry in Germany took a different turn in the eighteenth century, and Stahl was not unreasonably sceptical of the corpuscularians' premature attempts to effect a reduction of chemistry to physics. 'Mechanical Philosophy', he remarked,

> 'though it vaunts itself as capable of explaining everything most clearly, has applied itself rather presumptuously to the consideration of chemico-physical matters . . . it scratches the shell and surface of things and leaves the kernel untouched.'[8]

Explanations-in-principle would not do for Stahl; and his attitude was taken up again in the nineteenth century by William Whewell in his criticism of the atomic theories of his day.

Stahl led chemistry off into the 'jungle' of the phlogiston theory, but Lavoisier, who expelled phlogiston from the science, was equally sceptical about theories of matter. Thus he wrote:

> 'if, by the term *elements*, we mean to express those simple and indivisible atoms of which matter is composed, it is extremely probable we know nothing at all about them; but if we apply the term *elements* . . . to express our idea of the last point which analysis is capable of reaching, we must admit, as elements, all the substances into which we are capable, by any means, to reduce bodies by decomposition.'[9]

Any further discussions as to the nature of the elements could only be, according to Lavoisier, of a metaphysical nature. Here we have the notion of the 'chemical atom' which became widely accepted during the nineteenth century. Chemists did not need, and indeed should perhaps eschew, theories of matter; such problems should be left to physicists or, maybe, to philosophers. Chemists should be plain blunt men, keeping their feet on the ground, and in their science all that it was necessary to know was that some substances could not be further analysed. These were the elements; and the smallest units of them appearing in chemical reactions were the atoms. With the progress of the science, the list of elements might change as some yielded to more powerful techniques. In using the term 'atom' one was not making any statements about indivisibility; and indeed many English chemists used the formally absurd term 'compound atom' as the smallest unit of, for example, carbon dioxide. This 'atom' could certainly be decomposed into carbon and oxygen; but it would then have ceased to be carbon dioxide, and become something qualitatively different.

While many chemists were sceptical as to the value of the corpuscular philosophy in their science, preferring to leave open questions as to the ultimate nature of matter, some physicists were also dissatisfied with billiard-ball atomic theories. The most notable of these was the Jesuit, Roger Boscovich, whose name was for a long time almost forgotten but whose importance is now once again being realised. Boscovich described his theory as 'Natural Philosophy deduced from a single Law of Forces'. His atoms were mere mathematical points, the centres of forces which were at various distances attractive and repulsive, as in his diagram:

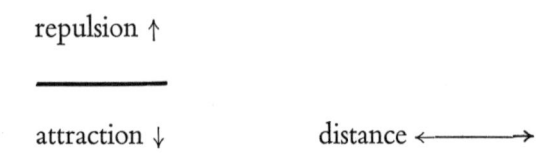

repulsion ↑

———————

attraction ↓   distance ←——→

The troughs represent attractive forces, and the peaks, repulsive. When two atoms approach one another, their combination will be stable if their distance apart corresponds to a trough on the diagram. The various troughs correspond to different combinations of different strengths; for example, chemical bonding, and the cohesion which holds solids and liquids together. The field of force of every atom extends through all space; at large distances it is attractive, and corresponds to gravitation, while at very short distances the repulsive force rises to infinity, so that two atoms cannot get near enough to occupy the same space. In Clerk Maxwell's illuminating metaphor, Boscovich atoms occupy space in the same way as soldiers occupy territory.

The major difficulty of the Boscovich atom was that it was an abstruse idea, much harder to visualise than a billiard-ball atom. Indeed Boscovich explicitly warned his readers of the dangers of simple atomic models:

> 'for the purpose of forming an idea of a point which is indivisible and non-extended, we cannot consider the ideas that we derive directly from the senses; but we must form our own idea of it by reflection'.[10]

This made the model seem arid and remote; and a further difficulty was that if matter were made up of mere mathematical points, it was hard to see how it could have mass. In Whewell's phrase, such a world would not be a material world.

The attractions of the model were mainly philosophical, though Faraday was able to make use of it in physics. The crudeness of the billiard-ball atom, with its associated action at a distance, was avoided. The mathematical points had a minimum of properties built into them, so a Boscovichean was not open to the reproach of seeking to explain, for example, the hardness of bodies by simply postulating hard atoms of which they are composed. But the exact dimensions of the peaks and troughs of the diagram were vague, and there was no method of calculating them.

All Boscovich atoms, more explicitly than Newtonian ones, were identical, since all were mere points. Nature resembles a library:

> 'Those books, so many in number and so different in character, are bodies, and those which belong to the different kingdoms are written, as it were, in different tongues.[11]

Chemical analysis discloses a

> 'large number of species of oils, earths, and salts from different bodies. Further analysis of these, like that of the words, would disclose the letters that are still less unlike one another; and finally, according to my theory, the little homogeneous points would be obtained. These, just as the little black circles formed the letters, would form the diverse particles of diverse bodies through diverse arrangement alone.'

The chemical atoms would be composed of a large number of Boscovich atoms, differently arranged in each substance.

Boscovich's theory seems to have aroused more interest in England than on the Continent; and in particular those associated with the Royal Institution, founded at the turn of the eighteenth and nineteenth centuries, appear to have become adherents. Notably both Davy and Faraday opposed the atomic theory of Dalton and proposed in its stead a Boscovichean atomic theory. For Davy the particular attraction of the model was that it demanded only one kind of atom, for Davy could not believe that every chemical element had its own distinct variety of indivisible atom. For Faraday the theory had the further advantage that it avoided action at a distance across empty space. We shall explore these views in later chapters.

Most chemists about 1800 remained corpuscularian in outlook, though their corpuscular beliefs had little impact on their chemical theory or practice. Before the days of chemical equations, and before the laws of chemical composition were clear, there was little or no cash value in an atomic theory for chemists. Lavoisier had indeed stressed the principle of conservation of matter in chemical changes, an important step towards chemical equations, and this principle found a ready support in an atomic theory. For if all matter be made up of indestructible atoms, of which the arrangements only were changed in chemical changes, then naturally matter must be conserved. But an atomic theory which merely accounted for conservation of matter, which is not after all incompatible with a continuum, was not very valuable or important. Such an atomic theory was not worth the trouble of confuting; and it seems that most scientists, and laymen too, were atomists without thinking very hard about the matter. Dalton sharpened the issue, gave more content to the atomic theory, and forced his contemporaries and successors to decide for or against it.

2

Mr Dalton and his Critics

DALTON presented his theory as an outgrowth of the corpuscular beliefs of his contemporaries. The chemical textbooks of about 1800, and the syllabuses of lectures delivered at the Royal Institution by Humphry Davy in 1802, and his predecessor Thomas Garnett in 1801, contain a reasonably consistent theory of matter, clearly the received opinion. Thus Davy wrote:

> 'the different bodies in nature are composed of particles or minute parts, individually imperceptible to the senses. When the particles are similar, the bodies they constitute are denominated simple, and when they are dissimilar, compound. The chemical phenomena result from the different arrangements of the particles of bodies; and the powers that produce these arrangements are repulsion, or the agency of heat, and attraction.'[12]

It was by no means evident that any simple bodies in this sense were known. Lavoisier had rejected any idea of defining chemical elements in terms of an atomic theory, and there was no reason to suppose that bodies which resisted analysis were composed of similar particles. According to these corpuscularians, there were two important kinds of forces at the particle level. The 'attraction of aggregation', sometimes called the 'attraction of cohesion', held like particles in proximity to one another in solids and liquids. When a chemical reaction took place, this 'attraction' was supposed to be overcome by the 'attraction of affinity' of the particles of the two different reacting substances. Affinities could be to some extent measured, or at least compared, and various tables of them were prepared in the latter part of the eighteenth century. But in general the theory could provide no more than explanations in principle of how chemical change was possible.

The theory was Newtonian, and authors were usually careful to point out, as Newton had in discussing gravitation, that they used the term 'attraction' in a descriptive and not an explanatory manner. Thus in his textbook Murray wrote:

> 'in the language of modern philosophy ["attraction"] is employed merely as an expression of the general fact, that the masses or particles of matter have a tendency, when left to themselves, to approach until they come into apparent contact.'[13]

Since some kind of belief in atoms was almost universal, Dalton's new theory was not treated by his contemporaries as an *atomic* theory at all, but as a statement of laws of chemical composition. In Dalton's day, it was recent knowledge that the elements combined together in fixed proportions by weight; and Dalton among others showed that they combined in definite and simple ratios. Dalton's proposals seemed to be that all the different chemical elements were composed of different sorts of indivisible atoms. All atoms of a given element, such as iron or oxygen, were identical. These atoms formed compounds by combining, in a manner Dalton did not explore, in simple ratios. If only one compound of a given pair of elements is known, then it is to be supposed 'binary', that is composed of two atoms, one of each component. Water was therefore made up of one atom of oxygen and one of hydrogen. If a number of compounds are known, as for example the series of oxides of nitrogen, then the simplest must be binary; the next ternary, two atoms of one element and one of the other; and so on. These simplicity rules produced some order, but although they had some basis in geometry they seemed to Dalton's contemporaries arbitrary and unscientific. And indeed they led to what we now know are the wrong formulae for water and confusion over the oxides of nitrogen.

Nevertheless, armed with these rules one can calculate the relative weights of atoms. Since one unit by weight of hydrogen combines with eight units by weight of oxygen to form water, if water is indeed binary, then the oxygen atom must weigh eight times as much as the hydrogen one. And if one takes the hydrogen atom as the basis of one's scale of relative weights, and assigns to it unit weight, then the atomic weight of oxygen will be eight units. Dalton did take hydrogen, the lightest element, as his unit; but many chemists, mindful of the very important role played by oxygen in the science, took that as their basis and assigned to it the weight 1, 10, or 100. This was a step of extreme importance in the history of chemistry, for much which had been qualitative could now become quantitative; but Dalton's contemporaries saw at once that this valuable step could be taken advantage of without accepting the atomic hypothesis upon which it appeared to be based.

This attitude is made very clear in the speech made by Davy in 1826 when, as President of the Royal Society, he presented a Royal Medal to Dalton. The praise seems somewhat grudging; but there seems no reason to suppose that Davy was jealous of Dalton or was doing anything worse than voicing, in a rather unpleasant way, misgivings about the theory which most of his audience would have shared. The geologist Gideon Mantell wrote of the occasion:

'Sir Humphrey Davey [*sic*], P.R.S., was in the Chair, and

made an eloquent and luminous speech on the merits of the celebrated individuals who were honored with the medals.'

It is worthy of note that the rules governing the award of the medals had been specially relaxed so that Dalton could be so honoured, even though his theory had by then been out for more than twenty years.

The medal was awarded, Davy said:

'for the development of the chemical theory of definite proportions, usually called the Atomic Theory . . . [*Dalton*, first laid down, clearly and numerically, the doctrine of multiples; and endeavoured to express, by simple numbers] the weights of the bodies believed elementary. His first views, from their boldness and peculiarity, met with but little attention; but they were discussed and supported by Drs Thomson and Wollaston; and the table of chemical equivalents of this last gentleman separates the practical part of the doctrine from the atomic or the hypothetical part, and is worthy of the profound views and philosophical acumen and accuracy of the celebrated author. . . . With respect to the weight or quantity in which the different elementary substances entered into union to form compounds, there was scarcely any distinct or accurate data. Persons whose names had high authority differed considerably in their statements of results; and statical chemistry, as it was taught in 1799, was obscure, vague and indefinite, not meriting the name of a science. To Mr Dalton belongs the distinction of first unequivocally calling the attention of philosophers to this important subject . . . thus making the statics of chemistry depend upon simple questions in subtraction or multiplication, and enabling the student to deduce an immense number of facts from a few well-authenticated, accurate, experimental results.'[14]

This passage gives the clues in our attempt to see how Dalton's theory was received. The acceptable part of it was the empirical part, laying down the laws and ratios according to which the elements combined together. Even this was not altogether acceptable until Wollaston and Thomson had, in their accurate researches on the oxalates, provided a sounder experimental basis than Dalton had himself been able to achieve. Besides this empirical portion, Dalton's theory seemed to contain hypotheses of a bold and extraordinary kind. The main value of the theory was that it saved much time which would otherwise have had to be spent on quantitative analyses.

It seems that Dalton's critics, amongst whom must be included most of his English contemporaries, may be divided into three broad groups. First come those whose hope was for a mathematical chemistry; a science like post-Newtonian astronomy, in which detailed calculations could be made by those skilful enough to handle the equations. Newton had been unable to discover the cause of gravity, but he had brought an enormous range of phenomena under a single law, enabling his successors to make detailed explanations and predictions in a wide field. Chemists should expect that the history of their subject would parallel that of astronomy. Dalton could not expect in such a story a position higher than that of Kepler, who had provided the laws of planetary motion which Newton later generalised. Kepler had indeed tried to explain the motions of the planets in terms of magnetic forces; but this hypothesis had soon sunk into oblivion. The same might be expected to happen to Dalton's atomic hypothesis.

The second major group included large numbers of chemists in Dalton's own day and in succeeding generations. They held that atoms were unnecessary theoretical entities. Nobody could prove the existence of atoms, and chemistry would make more progress if instead of postulating hypothetical and unobservable entities its practitioners sought the laws connecting observable phenomena. The most important of these positivists in Dalton's generation was William Hyde Wollaston.

The third group, of whom Davy was the most important, believed in the unity of matter. In the corpuscularian and Boscovichean tradition, they held that all matter was made up from identical atoms, differently arranged. Dalton's theory seemed opposed to this conception. According to Dalton, all atoms of iron, for example, were identical, and so were all atoms of hydrogen; and the implication was that the different kinds of atom were irreducibly different. Those who believed in the unity of matter continued to oppose the atomic theory throughout the nineteenth century; until in fact the theory was revised, in the light of physical discoveries, and a complex structure was assigned to the atom, about the turn of the century.

These three groups naturally overlapped to some extent; Davy, for example, seems from his writings to have attacked Dalton's theory on all three grounds. But the differences between their positions makes it worth while to examine the three groups of critics in turn.

Those who hoped to make chemistry mathematical believed that the corpuscular philosophy would somehow acquire the exactitude of Newtonian astronomy. One route towards this desirable end seemed to lie in the measurement of the strengths of the various affinities. Guyton, an associate of Lavoisier's, looking forward to a mathematical chemistry wrote:

> 'the phenomena of combinations produced or destroyed are not the result of occult qualities, but of a rupture of equilibrium determined by forces which afford the hope of admeasurement by computation'.[15]

And Pearson, who translated into English the new nomenclature of Lavoisier and Guyton, was also sanguine

> 'that the same certainty as in mathematics may hereafter be attained in chemistry'.

Others among Lavoisier's colleagues held similar views. The mathematical physicist Laplace had even done some calculations, which established that the attractions which appeared to exist at the microscopic and macroscopic levels—affinity, aggregation, and gravitation—might all be manifestations of one power, as Berthollet and others in the corpuscularian traditions had believed. But in order to produce gravitational forces of sufficient power to account for the phenomena of chemistry, the atoms would, according to Laplace's calculations, have to be almost unimaginably small and dense. Laplace believed that the time for such speculations had not yet come, since the laws of affinity were not yet known with sufficient exactness. He seems to have hoped that experiments on thermal decomposition might provide the data that were required. He declared:

> 'Some experiments already made afford us reason to hope that one day these laws will be perfectly known, and that then, by the application of analysis, the philosophy of terrestrial bodies may be brought to the same degree of perfection, which the discovery of universal gravitation has procured to astronomy.'[16]

To anybody looking for some kind of mathematical laws of chemistry, Dalton's codification of the laws of chemical combination would be useful, and his relative 'atomic' weights a step in the right direction. But his attempt to provide a causal explanation in terms of atoms would seem irrelevant and retrograde. The physics of falling bodies and of planets had made rapid progress when in the seventeenth century scientists stopped seeking for causal explanations only, and looked instead for the laws to which the phenomena conformed. Causal and mechanical explanations might have heuristic, expository, or even explanatory value, according to this way of thinking, but they must not be regarded as ends in themselves.

William Hyde Wollaston, the leading positivist among Dalton's antagonists, was a doctor who made his fortune by inventing a process for making platinum malleable. Earlier workers had only been able to isolate the metal in the form of a grey powder. Wollaston then devoted his time to research, abandoning his practice, and produced papers on a great range of topics; including the problems of fairy rings, and why the eyes in portraits follow one around. He became one of the most distinguished Fellows of the Royal Society, and was famous for his caution and the nicety of his analytical techniques. He had himself apparently been contemplating some kind of atomic theory, but had been forestalled by Dalton.

Nevertheless, it was Wollaston's analysis of the oxalates, published in the *Philosophical Transactions* of the Royal Society in 1808, which forced the chemical world to take the atomic theory, or at least the law of multiple proportions, seriously. In this paper Wollaston, distinguishing between laws of nature and the models used to account for them, remarked that it would be possible to treat atomic weights as mere arithmetical ratios, thereby avoiding any commitment to a theory of matter. This seemed to him an inadequate approach. He wrote:

> 'I am further inclined to think that when our views are sufficiently extended to enable us to reason with precision concerning the proportions of elementary atoms, we shall

find the arithmetical relation alone will not be sufficient to explain their mutual action, and that we shall be obliged to acquire a geometrical conception of their relative arrangement in all the three dimensions of solid extension.'[17]

In other words, to explain the chemical nature of a substance it would not be enough to know the proportions in which its component elements were present, but it would be necessary also to know how the atoms were arranged in space. A few years later, in 1813, he tried to execute this programme, and published a paper—following Hooke a century and a half before him—showing how various crystal structures could be built up from spherical or spheroidal particles. But, by 1814, he had abandoned all attempt to achieve an atomic geometry, and fell back on the arithmetical ratios which he had earlier thought would prove inadequate.

His paper, praised by Davy in the quotation above (p. 19), described a 'synoptic scale of chemical equivalents'. Instead of 'atomic weights', Wollaston spoke now of 'equivalents':[18] 'When we estimate the relative weights of equivalents, Mr Dalton conceives that we are estimating the aggregate weights of a given number of atoms.' Experiment revealed the proportions of elements in compounds, and the ratios in which elements combined. These latter values, the equivalents, were independent of all hypothesis. When, on the other hand, Dalton applied his simplicity rules and deduced that a compound was 'binary' or 'ternary', he went beyond the facts. No experiments could illuminate this dark region. Since it could never be known whether a given compound were really 'binary' or 'ternary', it was hopeless to seek true atomic weights. Equivalents were soundly based on facts and were just as useful to the chemist.

Most textbooks accepted this doctrine, and when they used the terms 'atom' or 'atomic weight' they made it clear that what was meant was 'equivalent'. 'Atom' continued to be used for its convenience; and in speaking of, and welcoming, the atomic theory, writers were careful to add that it should be accepted only in a form divested of all hypothesis. The success of the positivists, in terms of the wide acceptance of their attitude among chemists, during the next half-century, reveals the astonishing infertility of a specifically *atomic* theory in chemistry before the 1860s. As Wollaston had said, it would be necessary, before one began to think about the arrangements of atoms, to reason with precision concerning their proportions; and it was Cannizzaro's paper of 1860 which brought about this agreement on atomic weights.

Dalton could only explain the laws of constant, multiple, and reciprocal proportions; if elements were made up of identical atoms, and compounds of different atoms in fixed ratios, then naturally elements must react in definite ratios by weight. But it was possible to accept the laws of chemical combination as given, as brute facts, and to reject the atomic theory because of the high price in hypothesis which had to be paid for a small quantity of explanation. Hypothetical entities, or models, are acceptable in the sciences when their use leads to a wide range of phenomena being explained, and perhaps to predictions being made; or when they help in making discoveries, or in teaching the subject. That is, such models should have an explanatory, heuristic, or expository value. In the second decade of the nineteenth century, the atomic theory had none of these advantages in a very pronounced way. Only with the rise of structural chemistry in the 'sixties and 'seventies did they become generally apparent.

Davy agreed with Wollaston, but for different reasons. In his *Elements of Chemistry* of 1812 he used the term 'numbers representing the undecompounded bodies' where Dalton had written of the atomic weights of the elements. And his address of 1826, from which we quoted above, gave support to the positivists. But Davy was not himself a positivist, and in his writings he made few efforts to eschew hypotheses; rather, he was a free and easy user of them, taking them up and dropping them as he went along. He had a distinctly speculative turn of mind which fortunately he never managed to curb completely. He rejected the atomic theory because, in a sense, it was not hypothetical enough. For Dalton appeared to postulate that the various elements were all composed of irreducibly-different atoms; while one of the beliefs to which Davy consistently clung was that the elements were all composed of the same kind of atoms, in different arrangements.

Davy was, at the time when Dalton's theory began to arouse attention, at the height of his powers. By 1808 he was easily the best-known chemist in England. Coming from a humble background, he had made his name with his researches on laughing gas, and had been appointed Professor of Chemistry at the Royal Institution in his early twenties. His lectures attracted large and fashionable audiences, ensuring the economic survival of the Royal Institution; and he was in great demand at dinner parties. Davy had begun to apply the newly discovered electric battery to chemical analysis, and in the issue of the *Philosophical Transactions* which contained the papers by Wollaston and Thomson supporting the atomic theory, Davy announced the discovery of the new metallic elements sodium and potassium. These light and highly reactive substances—potassium floats on water and decomposes it, setting fire to the hydrogen evolved—caused a sensation in the chemical world, and lent themselves to dramatic lecture demonstrations. Davy

> 'ventured to conclude from the general principles on which the phenomena were capable of being explained, that the new methods of investigation promised to lead to a more intimate knowledge than had hitherto been obtained concerning the true elements of bodies'.[19]

Davy had expected that potash and soda, which had hitherto resisted analysis, must be decomposable, like all Lavoisier's elements, by some agent. Electricity had indeed done the trick; and Davy was soon able to isolate other metals in the same way. He thus proved that soda and potash were not elements, but were the oxides of the new metals. This left the total number of known elements unchanged; the metals just replaced the oxides on the list. But Davy does not seem to have found this discouraging. Ammonia, known to be a compound of nitrogen and hydrogen, formed salts extremely like those of the new metals. These bodies must therefore be compounds too; and some tool of even greater power than the electric battery was required in order to break down these metals, and indeed all the metals, into their true elements.

His brother later told the story of how Davy was converted to a belief in the atomic theory by Wollaston and a friend. According to John Davy, Humphry's views were 'a modification of those of Mr Dalton—the same in regard to fact, stripped of all speculation'.[20] But all that this means is that Davy was persuaded that elements react in definite and constant and multiple proportions; Dalton's atomic theory he consistently opposed. And the main reason for his opposition remained his conviction that the number of real elements must be extremely small.

In 1809 Davy, in a lecture at the Royal Institution, explained that since both elements and compounds were known to react in definite proportions, we cannot, from the laws of combination, deduce anything about their status. The very different chemical properties of the elements, again, tell us nothing about their ultimate irreducibility, for many compounds of the same two elements exist which differ very widely in their properties. Davy's examples were the oxides of nitrogen; air (which he believed to be a compound), laughing gas, and nitric oxide, he declared:

'though composed of the same elements, yet are themselves capable of combination as if they consisted of distinct materials. And these facts, combined with analogous facts relating to the compounds of hydrogen, carbon, and oxygen, render it probable that substances which we at present conceive to consist of different species of matter may ultimately be referred to different proportions of similar species, and in this way the science of the composition of bodies may be materially simplified.'[21]

This argument can do no more than show that it is not unreasonable to suppose that different proportions and arrangements of a few real elements could generate all the apparent elements, the bodies which resist analysis. The resemblance between compounds or radicals on the one hand, and elements on the other, was emphasised throughout the nineteenth century by those who wished to deny a special status to the latter.

If Davy's reasonings established that it was possible that the elements were complex, it would still be open to anyone to ask why one should wish to believe that they are. Davy's mentor, Thomas Beddoes, had written:

'Whether to create a diversified system of bodies out of one, or out of a few or many elements, imply most wisdom or power, is a question which different persons would decide according to their various taste in world-making.'[22]

And in a recent book Rom Harré has shown that it is not possible to lay down criteria for simplicity which everyone will accept. In the case of the elements, one had either to choose with Dalton a world composed of a large number of elements, whose arrangements in compounds were relatively simple; or to believe in very few elements of which the atoms were extremely simple in properties, but were arranged in complex and various ways to produce all the richness and variety we encounter in the world.

Davy had no doubt which to choose. Simplicity of elements rather than simplicity of relations was what he looked for. A chemistry based on few elements would lend itself more easily to quantification and to mathematical treatment; there would result

> 'a new, a simple and a grand philosophy.... From the combination of different quantities of two or three species of ponderable matter, we might conceive all the diversity of material substances to owe their constitution, and as the electrical energies of bodies are capable of being measured, and as these are correspondent to their chemical attractions, so the laws of affinity may be subjected to the forms of the mathematical sciences, and the possible results of new arrangements of matter become the objects of calculation.'[23]

This resembles what Laplace had written, but Laplace had worked with Lavoisier on heats of reaction and hoped that chemistry would be quantified by further similar experiments, whereas Davy's researches had been in the field of electricity, a science then rapidly becoming mathematical. Davy's researches had demonstrated that some chemical bonds were electrical in nature; and Davy believed all were.

In the sentence just quoted, Davy is clearly looking forward to a distant epoch whose beginning we are privileged to see; the time when it will be possible to predict chemical properties from electronic configurations. This was far more ambitious than anything Dalton could have hoped to achieve with his atomic theory. Indeed such speculations had little to do with nineteenth-century chemistry at all; but Davy was too much a physicist at heart to ignore theories of matter and rest content with 'chemical atoms'.

Though one reason for believing in the unity of matter was that it seemed to bring closer the prospect of a mathematical chemistry based on electrical researches, the main attraction of the hypothesis seems to have been, for Davy at least, metaphysical. Davy's lecture of 1809 ends with a splendid passage:

> 'From the past progress of the human mind, we have a right to reason concerning its future progress. And on this ground a high degree of perfection may be expected in chemical philosophy. Whoever compares the complication of the systems which have been hitherto adopted, and the multitude, as it were, of insignificant elements, with the usual simplicity and grandeur of nature, will surely not adopt the opinion, that the highest methods of our science are already attained; or that events so harmonious as those of the external world, should depend upon such complex and various combinations of numerous and different materials.'

When this passage appeared in slightly modified form in Davy's *Elements of Chemistry*, it was hailed by a reviewer as 'pious yet sublime'.

In 1811, in another lecture, Davy declared directly that he was opposed to Dalton's atomic theory because it seemed to deny the divisibility of the particles of which the chemical elements were composed. In fact, as we shall see, Davy seems to have taken Dalton's theory more seriously as a theory of matter than Dalton did himself. Davy believed that the true exlpanation of definite proportions would be found not in the atomic theory but in 'the identity of the matters really acting upon one another'. Despite his views, Davy was not infrequently given credit in textbooks and reviews for helping to confirm and extend the atomic theory. This can only have happened because to most contemporaries of Dalton and Davy the term 'atomic theory' meant no more than 'the laws of chemical composition'.

Dalton, faced with such an array of antagonists, tried to counter-attack. One of the problems in understanding his views is that the atomic theory was not suddenly produced all at one moment in its final form. Indeed it has been remarked that Dalton only became a Daltonian under the pressures of controversy over a number of years following the publication of his theory. There seems to be little doubt that Dalton developed his theory in the first place in the course of his work on mixed gases, and then gradually applied it in chemistry, a science with which he was not at first particularly concerned although he took an interest in it. His theory therefore developed from being primarily a theory of matter into 'chemical atomism' over a period of years.

Dalton had read Newton, and had copied out Query 31 from the *Opticks* into his notebook. Newton's work on gravity showed, according to Dalton, that: 'however guarded we should be, not to let a theory or hypothesis, contradicted by experiment, mislead us; yet it is highly expedient to form some previous notion of the objects we are about, in order to direct us into some trains of enquiry'.[24] In other words, hypotheses had an heuristic value. Davy would have agreed with this; he often wrote that the only value of hypotheses was to lead to the performance of new experiments, and he was prepared to discard hypotheses with some abandon. But such hypotheses would have had to be candidates for truth in a way that the atomic theory could not, in his view, be. And Dalton's claims went further than that, anyway. In a letter to Berzelius, he claimed that without an atomic theory definite proportions would be 'mysterious'. Comparing chemistry with astronomy, he cast himself as both Kepler and Newton. Definite proportions were 'like the mystical ratios of Kepler, which Newton so happily elucidated'. Hard thinking on the matter could only terminate, he conceived, 'in the arrangements of particles as exhibited in the diagrams in the *New System of Chemistry*'. Dalton's highly theoretical and *a priori* cast of mind is revealed by his remarks on Gay-Lussac's investigation of gaseous reactions; that we should not 'be led to adopt these analyses till some reason be discovered for them'.

The argument that the laws of chemical combination would be mysterious without an atomic theory was not really very powerful. It was very like what Newton's Cartesian critics had said; that the law of gravity was a perpetual miracle unless one supposed that there were great swirling whirlpools of æther to explain it. Before the atomic theory could provide testable explanations and predictions, it was a matter of taste whether one preferred to 'explain' the laws of combination in terms of an unverifiable atomic theory, or simply to accept the laws as given. Dalton after all could only explain definite and multiple proportions, and even here he only succeeded in pushing the explanation back one stage without adding anything to it. This situation reminds us of the old story that the earth was supported by an elephant, and the elephant by a tortoise; but what held the tortoise up was left vague. All this changed when the range of the atomic theory was enormously extended, particularly by Kekulé and his school, in the second half of the nineteenth century.

Dalton had little success in persuading the positivists that his theory swept away mystery, as the unanimous presentation of Wollaston's views in the standard textbooks shows; and he did not fare any better in dealing with adherents to the theory of the unity of matter. Dalton never seems to have been able to make up his own mind whether he intended his theory as a theory of matter, or as a theory of chemical atoms. Thus he spoke of 'simple' and 'compound' atoms, using the term 'atom' where we should write 'molecule'. His rather muddled views are illustrated in a lecture of 1810, which is clearly intended as a refutation of Davy.

'It has been imagined by some philosophers that all matter, however unlike, is probably the same thing; and that the great variety of its appearances arises from certain powers communicated to it, and from the variety of combinations and arrangements of which it is susceptible. From the notes I borrowed from Newton in the last lecture, this does not seem to have been his idea. Neither is it mine. I should apprehend there are a considerable number of what may be properly called *elementary* principles, which never can be metamorphosed, one into another, by any power we can control. We ought, however, to avail ourselves of every means to reduce the number of bodies or principles of this appearance as much as possible; and after all we may not know what elements are absolutely indecomposable, and what are refractory, because we do not apply the proper means for their reduction. We have already observed that all *atoms of the same kind*, whether simple or compound, must necessarily be conceived to be alike in shape, weight, and every other particular.'[25]

There are several interesting points which emerge from this long quotation, full of hedging. The first sentence is directed at Davy; but having declared that his taste in world-making would incline him to use several different kinds of element, Dalton moves away rapidly from a theory of matter. If atoms can be either simple or compound, then they are quite different in conception from the absolutely hard and unsplittable particles of the corpuscular philosophy. Indeed, when atoms are merely such units in chemical reactions, they differ hardly at all from equivalents. This weakening meant that the atomic theory, no longer a theory of matter, could not bring chemistry and physics together. A. W. Williamson, defending the atomic theory before the Chemical Society in London in 1869, was in the tradition of Lavoisier, and of the Dalton of the last sentence of the above quotation, when he disavowed interest in the real nature of the atoms.

In the years after about 1820 chemistry, in Britain at least, passed through a period of consolidation, a period in which atomic weights or equivalents—the two terms were used interchangeably—were determined with increasing accuracy. Chemists no longer looked for the one basic law of nature governing all phenomena, or similar chimeras. And public attention shifted from chemistry to geology. As Whewell wrote of a different science:

> 'We may observe also that we have now described the period of most extensive activity and interest . . . This naturally occurred while the general notions and laws of these phenomena were becoming, and were not yet become, fixed and clear. At such a period, a large and popular circle of spectators and amateurs feel themselves nearly upon a level, in the value of their trials and speculations, with more profound thinkers: at a later period, when the subject is become a science, that is, a study in which all must be left far behind who do not come to it with disciplined, informed, and logical minds, the cultivators are more few, and the shout of applause less tumultuous and less loud. We may add too, that the experiments, which are the most striking to the senses, lose much of their impressiveness with their novelty.'[26]

Never again would a chemical lecturer attract, as Davy had, vast and fashionable audiences, to be staggered at the demonstrations of the properties of the alkali metals. One can sympathise with those who felt that in the transition from a qualitative to a quantitative science the glory had passed away.

Andrew Ure, otherwise known as an apologist for the factory system, in a review of a book by Thomas Thomson full of atomic weight determinations, deplored the change.[27] The body and soul of chemistry, he declared, did not reside in a knowledge of the combining weights of substances. The characteristics of chemical genius were to discover new elementary bodies, new qualities, and to arrange under general laws the phenomena of corpuscular action. Qualitative research should be preferred before quantitative; atomic weight determinations are a banausic activity. Ure nostalgically appealed for a return to the spirit of Davy, Berthollet, Wollaston, and Gay-Lussac, who had researched into the powers that modify matter.

This appeal went unheard, or at any rate unanswered, and the glamour with which Davy had invested the science did not survive his death in 1829. Instead the subject became a byword for its plodding methods. Sir John Herschel hoped that this experimental phase would be followed by a period in which it would become a mathematical, deductive science. Because the laws of chemistry had been arrived at by simple enumerative induction, they were in his view mere generalisations, and yielded no important discoveries. Chemical theories were

> 'for the most part, of that generally intelligible and readily applicable kind, which demand no intense concentration of thought, and lead to no profound mathematical researches'.

Because of the complexity of the subject, the axioms of chemistry, which would form the basis of a deductive science, were still unknown. Though Herschel thought the atomic theory a step of great importance, in fact he meant 'the law of definite proportions, which is the same thing presented in a form divested of all hypothesis'.[28]

Dalton was the Kepler of chemistry, which still awaited its Newton. This was a period when the mathematical physicists were particularly arrogant; but with regard to the atomic theory, their position was by no means unjustifiable. And indeed throughout the century those who tried to make chemistry a mathematical science tended to have little to do with atomism, as the examples of Brodie and Ostwald will show.

In fact all three lines of opposition to Dalton's theory persisted through most of the century. And in the next chapter we shall look at the views of Davy and Faraday, who sought to apply the Boscovich atom in chemistry and physics; and of Whewell, who opposed all atomic explanations.

3

Boscovicheans and Sceptics

So far we have met Davy as a critic of Dalton's theory, chiefly on the grounds that it involved atoms of too many different kinds. Put another way, this criticism comes down to saying that the atoms were not adequately simple in their properties. Epicurus had written that one must believe the atoms to possess none of the qualities belonging to things, except shape, weight, and size; and seventeenth-century atomists had followed him. Dalton's atoms, on the other hand, must have had far more qualities than these; in fact, being chemical atoms, they must have been supposed to possess all the chemical properties of the large bodies composed of them. The differences in chemical properties between sodium and potassium would have been explained by a Daltonian as entirely due to the differences between their atoms. With such atoms, therefore, one could not hope really to explain chemical properties; for these were simply built into each atom. It is in fact only with increasing knowledge of atomic structure and sub-atomic particles that present-day chemists have been able to provide any such explanations.

An atomic theory which does not explain the complex properties of bodies in terms of the different arrangements of atoms with much simpler properties is, to say the least, unsatisfactory as a theory of matter. And the opposition of so many people to such a theory is not surprising. In this chapter we shall see how Davy and Faraday tried to apply a theory that of Boscovich, in which the atoms were almost destitute of all properties, since they were reduced to mere mathematical points. In Clerk Maxwell's phrase, Boscovich's theory represents the purest monadism. William Whewell refused to accept even this kind of atomism. Believing that all atomic theories involved the explanation of the properties of things by postulating atoms composing them, endowed with the same properties; and that this kind of explanation was delusive and highly disreputable; he would have nothing whatever to do with such theories.

Davy began as a corpuscularian, as his syllabus of lectures for 1802 shows; and even at that date he stated clearly that there was a distinction between the elements as defined by Lavoisier and the simple bodies of the corpuscularians. At the Royal Institution his predecessor, Thomas Garnett, had discussed the theory of Boscovich; the *Outline* of his lectures of 1801—by which time Davy had arrived as his assistant—contain the following summary:

> 'Our ideas of impenetrability less certain than we have suspected. It is highly probable that the tangible particles of matter are not in contact, but are connected by mechanical forces, which, like gravity, act at a distance. Theory of Father Boscovich.'[29]

In his chemistry lectures, Garnett seems to have adhered quite closely to the ordinary corpuscularian view, and spoke of the attractions of aggregation and affinity in the usual way.

It was in fact the greatest problem in connection with the Boscovich atom that despite its philosophical attractions it was very difficult to incorporate into chemistry. As Charles Daubeny, Professor of Chemistry at Oxford, wrote in 1831, Boscovich's theory was ingenious, and skilfully evaded the difficulties which beset all who endeavour to assign limits to the divisibility of matter. But a world of Boscovich atoms would not be a real world, for they had too few properties; and 'the mass of mankind will be glad to escape from such obscure and abstract speculations'.[30] Instead of bothering with mathematics, most people would prefer to ask the question whether or not it was sound philosophy to admit atoms in a physical sense.

Davy must have come across Boscovich's theory either in Garnett's lectures or from one of the many authors of the time who mentioned it. In 1801, for example, the *Philosophical Magazine* published a short biographical memoir of Boscovich; and Priestley had earlier made his work well known. In his *Elements of Chemistry* in 1812, Davy wrote of

> 'the sublime chemical speculation, sanctioned by the authority of Hooke, Newton, and Boscovich, [which] must not be confounded with the ideas advanced by the alchemists'.[31]

The 'speculation' was that the same ponderable matter, in different electrical states or different arrangements, might constitute bodies with different chemical properties. By 1815, on his first Continental journey, with Faraday accompanying him and Lady Davy, he had become a convinced Boscovichean. And shortly before his death he had been working on the problem of explaining chemical phenomena in terms of phenoma using Boscovich atoms.

As a legacy to younger scientists, Davy wrote a book of dialogues, with the title *Consolations in Travel, or the last Days of a Philosopher*. This work, which was published posthumously, abounds in speculations and is fascinating to read. But unfortunately the dialogue on atoms was unfinished at his death, and therefore did not appear until Davy's *Collected*

Works were published in 1839-40. Cuvier wrote of the *Consolations* in his *Eloge* of Davy, that

> 'Once escaped from the laboratory, he had resumed the tranquil reveries and sublime thoughts which had formed the delight of his youth; it was in some measure the work of a dying Plato.'

In the dialogue on atoms,[32] a wise sceptical natural philosopher, the Unknown, expounds a theory of matter based upon Boscovich; with the chemical elements appearing as stable stopping places in the building up of chemical compounds from the point atoms. The Unknown began with an operational, theory-free definition of an element, based on Lavoisier's but with all hypothesis eradicated:

> 'I cannot demonstrate to you what are the true elements of things; but I can exhibit to you those substances, which, as we cannot decompose them, are elementary for us: . . . there is every reason to suppose that our powers of chemical decomposition have not yet reached their ultimatum; yet in the operations of nature, as well as in those of art, certain substances appear unchangeable. . . . The test of a body being indecomposeable is that in all chemical changes it increases in weight.'

It has been contended that 'element' must be a theory-loaded term, but it seems that Davy has here produced a definition which contains no theoretical concepts. The same definition was used by Ostwald about a century later; but his reasons for choosing it were somewhat different from Davy's. Ostwald would have nothing to do with atomic theories, and therefore an operational definition of an element was as far as one could hope to go; anything further that might be said about elements would be hypothesis, and belong to the realm of metaphysics. Davy, on the other hand, held an atomic theory of a sort—the theory of Boscovich—and was perfectly prepared to speculate about the ultimate structure of matter. So for him, the definition was required to delimit that class of bodies which all seemed to share the same status, and played an important role in chemistry. He might have hoped that mathematical progress would make it possible to drop the operational definition of 'element' and replace it by some definition in terms of configurations of Boscovich atoms; a step not unlike what has in fact happened, for in present-day chemistry 'element' is not defined as Lavoisier defined it but in terms of the number of protons in the nucleus of its atom.

In answer to questions about how differences in atomic weights are to be explained, the Unknown replied that such queries

> 'cannot be answered except by conjectures. At some time possibly we may be able to solve them by an hypothesis which will satisfactorily explain the chemical phenomena; but as we can never see the elementary particles of bodies, our reasoning upon them must be founded upon analogies from mechanics, and the idea that small indivisible particles follow the same laws of motion as the masses that they compose.'

Davy was always a firm believer in the importance of analogy in scientific reasoning. One of the characters in the dialogue takes up the Unknown at this point, remarking that lines are composed of points, and white light from coloured particles. The parts need not resemble the whole; the analogy between microparticles and masses need not be complete.

The Unknown denied the charge that he supposed, like a

Daltonian, that atoms had all the properties of the bodies composed of them:

> 'You mistake me if you suppose I have adopted a system like that of the Homooia of Anaxagoras, and that I suppose the elements to be physical molecules endowed with the properties of the bodies we believe to be undecomposeable. On the contrary, I neither suppose in them figure nor colour . . . I consider them, with Boscovich, merely as points possessing weight and attractive and repulsive powers; and composing according to the circumstances of their arrangements either spherules or regular solids, and capable of assuming either one form or the other.'

In liquids and gases the molecules would be spherical, and in solids they would be arranged in a regular way.

Now the Unknown proceeded to link this theory to the corpuscular philosophy. The molecules which we cannot decompose, the elements, are spherical; and that spheres could be packed so as to form regular solids Wollaston had shown. Further one must suppose that these molecules 'have certain attractive and repulsive powers which correspond to positive and negative electricity', in order to explain the formation of compounds. Davy's electrical work twenty years before had convinced him that the attraction of affinity was electrical in nature.

The Unknown tried to argue that all this was not mere speculation unsupported by experiment. In its favour he produced four facts: the first, that all bodies could be melted, and in the fluid state the particles must be supposed spherical to allow sufficient freedom of motion. The second was that all bodies were capable of forming regular crystals on cooling; and the third that all crystals 'present regular electrical poles'. The fourth fact was that the elements of bodies could be separated from one another by electricity. This belief in the importance of electricity Davy had shown in a lecture in 1810 when he had remarked:

> 'The voltaic battery was as an alarm bell to experimenters in every part of Europe; and it served no less for demonstrating new properties in electricity, and for establishing the laws of this science, than as an instrument of discovery in other branches of knowledge, exhibiting relations between subjects before apparently without connection, and serving as a bond of unity between chemical and physical philosophy.'

Now at last Davy had been able to sketch a theory of matter incorporating the discoveries of this branch of science.

Davy's Boscovichean theory could account for his four facts, but it could not be said that he had succeeded in showing that the theory was necessary for the explanation of the phenomena of chemistry; and his championship of it, especially since it was omitted from the very successful *Consolations in Travel*, was probably not even very widely known. Certainly no other chemists seem to have followed him in speculating along these lines. Pearce Williams has argued interestingly that the researches of Davy and Faraday on the liquefaction of gases by heating in sealed tubes, were based on their belief in Boscovich atoms. Such treatment could not possibly be expected to make billiard-ball atoms cohere; but Boscovich atoms might reach a trough on their curves and coalesce. Other chemical researches of Faraday become more readily intelligible in terms of his belief in point atoms; but other chemists, more down to earth perhaps, do not appear to have followed him.

We do not, it is true, usually think of Faraday as a chemist, but electricity was in the early nineteenth century treated as a branch of chemistry, and Faraday did also perform a number of researches which today would be counted as purely chemical. He never held office in the Physical Section of the British Association, but was President of the Chemical Section in 1837 and 1846. In some early chemical lectures, given when he was very much under Davy's influence, he presented views on the nature of the chemical elements in language very like Davy's; and to the end of his life he persisted in the conviction that matter could not be composed of many different kinds of elements.

In a lecture of 1816 Faraday gave it as the received opinion that light was composed of small particles, octahedral in shape, and went on to connect this view with speculations on the nature of matter. Newton, and Davy too, had believed that ordinary matter could be converted into 'radiant matter', or light, and that indeed this process was going on all the time. If this be assumed, said Faraday:

> 'We must also assume the simplicity of matter; for it would follow that all the variety of substances with which we are acquainted could be converted into one of three kinds of radiant matter, which again may differ from each other only in the size of their particles or their form. The properties of known bodies would then be supposed to arise from the varied arrangements of their ultimate atoms, and belong to substances only as long as their compound nature existed; and thus variety of matter and variety of properties would be found co-essential. The simplicity of such a system is singularly beautiful, the idea grand, and worthy of Newton's approbation. It was what the ancients believed, and it may be what a future race will realise.'[33]

The mysterious three kinds of radiant matter seem to have been a somewhat arbitrary number; for all sorts of radiation were included under this name.

BOSCOVICHEANS AND SCEPTICS

In 1819 Faraday returned to this idea, and made the suggestive remark that 'radiant matter' might be regarded as a fourth state of matter. As bodies are heated and change their state from solid to liquid, and then to vapour, so their physical properties become simpler and less differentiated. Gases have far more properties in common than do solids or liquids. In the transition from the gaseous to the fourth, radiant state more individual physical properties would disappear; so the properties of this state would be very simple, and there would probably be only three different kinds of radiant matter. Crookes much later took up this idea and applied to the cathode rays the name 'fourth state of matter'.

Faraday particularly deplored the increase in the number of the metallic elements, which chemists, following the lead of Davy and Wollaston, kept discovering. In 1818 he spoke of the way in which rocks, soil, sand, and salts had all been analysed and had yielded up metals. As Chenevix, the Copley medallist of 1803, had written:

'In a more enlightened period, we have extended our enquiries and multiplied the number of the elements; the last task will be to simplify; and by a closer observation of nature, to learn from what a small store of primitive materials all that we behold and wonder at was created.'[34]

Chenevix unfortunately believed that he had achieved such a simplification, in erroneously proving palladium to be a compound of platinum; but this was the programme which Davy attempted to carry out, with more competence but with no greater success. Faraday eloquently summoned his contemporaries:

'To decompose the metals, then, to reform them, to change them from one into another, and to realise the once absurd notion of transmutation, these are the problems now given to the chemist for solution. Let none start at the difficult task, and think the means far beyond him; every thing may be gained by energy and perseverence.'[75]

The story is well known, of how Faraday, then an old man, congratulated Crookes on the discovery of thallium, but remarked that to decompose the elements would be even better than to discover new ones.

Since the appearance of Pearce Williams' extremely important biography of Faraday, it is not necessary to labour to prove that Faraday was a convinced Boscovichean. But no papers explicitly avowing adhesion to Boscovich's theory were published by him until the 1840s; until after Davy's dialogue on the atomic theory had appeared in print. Instead there was an odd paper on limits of vaporisation, which provided an argument in favour of the view that the world was made up of atoms. The paper appeared at about the same time as one by Wollaston, seeking to establish the atomic nature of the Earth's atmosphere from astronomical observations. It is very odd to find these two men, both of them opponents of Dalton's theory, publishing these papers which seemed to support it; and odd too that although these papers aroused considerable interest it was some time before the faultiness of their reasoning was established.[36]

Wollaston's argument was that if the Earth's atmosphere were a continuum, it would shade off to infinity and not have a sharp boundary. If it had no boundary, it would be attracted by other heavenly bodies, and all the planets would have atmospheres. If on the other hand it were atomic, then the atoms on the boundary would be exactly balanced between the gravitational attraction of the Earth and the repulsion of other atoms. The boundary would therefore be sharp; no atoms would escape to other planets, and so other heavenly bodies need not share our atmosphere. Astronomical observations showed that other planets lacked atmospheres; therefore our atmosphere must be atomic. Abandoning his coolly positivist position, Wollaston suggested that these atoms might be those postulated by Dalton. Faraday's argument was similar; that if bodies were made up of atoms, then there would be a limit to vaporisation, a temperature below which no evaporation could take place because the repulsive force of heat would be less than the attractive force of aggregation. In a continuum, Faraday believed, there would be no such sharp limits. A limit to vaporisation is observed; so atoms must exist.

Both these arguments, we should notice, presuppose a highly static theory of gases; if atoms in a gas were rushing about with different velocities, as the kinetic theory supposes, the arguments would collapse. This was not seized upon by contemporaries—dynamic ideas only entered the science later—and there was general satisfaction with the arguments, which found their way into the textbooks. Damning criticism came from Whewell in 1839, seventeen years after Wollaston's paper had appeared. He showed that an equation could be devised for a continuous atmosphere which would have a sharp limit, and not fade away to zero density asymptotically as Wollaston had imagined it must.[37] Wollaston's argument therefore did not prove anything. And in the same way, Faraday's observations proved explicable in terms of the Phase Rule, which makes no assumptions as to the nature of matter, and requires none. A little later than Whewell, George Wilson remarked that even if one accepted the arguments as Wollaston [and Faraday] had originally proposed them, they only established the existence of molecules; not that these particles were unsplittable atoms. Wilson was not a believer in the atomic theory.

This episode does not seem very important in Faraday's thought about atoms, coming in a fallow period; and his biographer of 1898, when adulation of Dalton was at its height, thought it a flaw in Faraday's character that he failed to appreciate Dalton's greatness, and thought him and his atomic theory over-rated. As far as chemistry was concerned Faraday, like Davy, took up a positivist position: 'the words definite proportions, equivalents, primes, &c . . . fully express all the *facts* of what is usually called the atomic theory in chemistry'.[38] In chemistry, indeed, Faraday's thinking on atoms was almost identical to that of Davy; he believed that there could not be many different kinds of genuine elements, he was sceptical of Daltonian atomism, and he allowed the theory of point atoms to guide him.

His electrical researches were also inspired by his Boscovichean beliefs, but it was not until the 1840s that he permitted these ideas to appear in print. In 1844 there appeared the paper: 'A Speculation touching Electric Conduction and the Nature of Matter',[39] in which Faraday explored the relation between the structure of matter and its electrical conductivity. In the second paragraph of this essay, Faraday described the generally-held atomic theory of the day; and then proceeded to show how inadequate it was to account for the phenomena:

> 'The view of the atomic constitution of matter which I think is most prevalent, is that which considers the atom as something material having a certain volume, upon which those powers were impressed at the creation, which have given it, from that time to the present, the capability of constituting, when many atoms are congregated together into groups, the different substances whose effects and properties we observe. These, though grouped and held together by their powers, do not touch each other, but have intervening space, otherwise pressure or cold could not make a body contract into a smaller bulk, nor heat or tension make it larger; in liquids these atoms or particles

are free to move about one another, and in vapours or gases they are also present, but removed very much further apart, though still related to each other by their powers.'

The theory here described is very similar to that in the Queries in Newton's *Opticks*. The caloric fluid which filled the space between the atoms for Dalton and his contemporaries had by the 1840s disappeared; and the atoms were supposed to act upon each other at a distance across void space, a notion which Faraday found extremely distasteful. He added that while the term 'atom' was often used with the intention of expressing mere facts, he had 'not yet found a mind that did habitually separate it from its accompanying temptations'. Terms like 'equivalent' had not caught on, in Faraday's view, because they were not expressive enough; they did not express hypothesis as well as fact. He was, no doubt, right; those who wished, rightly or wrongly, to form a picture of a chemical reaction would have favoured an atomic theory above a mere statement of facts. But for Faraday, as for Davy and Wollaston, it was vital to distinguish fact from theory.

Faraday next proceeded to dig deeper into this received opinion. Matter is composed, on this view, of a kind of three-dimensional net of space, with particles where there would be meshes in the net. In a non-conductor, since the particles are not in contact, and a web of space runs through the whole, this space must be the insulator; for if it were a conductor then the substance would be one. In a conductor such as a metal, on the other hand, the only continuous part is again the space, and the atoms not only do not touch but are some distance apart. So space must be a conductor. We are left with a most unpleasant paradox.

Further, Faraday divided the density of various metals by their atomic weights, to find the proportional numbers of atoms in a given volume of each metal. And he found that these numbers bore no relation whatever to their conducting powers. Even odder, when he performed this calculation for potassium metal, which is very light, and potassium hydroxide, he found that the latter substance actually contained more atoms of potassium in a given volume than did potassium metal. And yet the metal is a good conductor, while dry potash does not conduct electricity at all. In potassium, then, the atoms must be very far apart; and so the space between them must conduct excellently.

Despite these inconsistencies, Faraday admitted that the phenomena of crystallisation, of chemistry, and of physics, pointed to the existence of some kind of centres of force. It was the notion of discrete particles with space between them which he found hard to swallow. As some assumptions were necessary, the safest course was to keep them to a minimum, and therefore to follow Boscovich, whose atoms: 'are mere centres of forces or powers, not particles of matter, in which the powers themselves reside'. If, in the ordinary view of atoms, we call the particle of matter away from the powers a, and the system of powers or forces in and around it m, then in Boscovich's theory a disappears, or is a mere mathematical point, whilst in the usual notion it is a little unchangeable, impenetrable piece of matter, and m is an atmosphere of force grouped around it.

As far as crystallography, definite proportions, and magnetism were concerned, it made little difference, according to Faraday, whether one adhered to Daltonian or to Boscovichean atoms; but the latter had the advantage in explaining electrical conductivity, the nature of light, the manner in which bodies combine, and the effects of forces on matter. All the properties of things in fact depend on the m rather than the a, which has no powers at all; so, paradoxically, the m is the matter of the atom. The difference between a supposed hard little particle and the powers around it Faraday found unimaginable:

> 'To my mind, therefore the *a* or nucleus vanishes, and the substance consists of the powers or *m*; and indeed what notion can we form of the nucleus independent of its powers?'

To think of matter devoid of, or independent of, powers is impossible.

A world of Boscovich atoms is a full world; in a gas the atoms touch one another just as truly as in a solid or liquid, though the centres of force differ in their distances. The problem which faced Daltonians in explaining conduction and insulation therefore disappears; there is no void space, and therefore no question of space being a conductor or insulator. The atoms are highly elastic, instead of hard, and compression of a body does result in compression of its atoms. The 'shape' of an atom now means the 'disposition and relative intensity of the forces'. The impenetrability of atoms disappears, and chemical combination is seen in a very different light:

> 'If we suppose an atom of oxygen and an atom of potassium about to combine and produce potash, the hypothesis of solid unchangeable impenetrable atoms places these two particles side by side in a position easily, because mechanically, imagined, and not unfrequently represented; but if these two atoms be centres of power they will mutually penetrate to the very centres, thus forming one atom or molecule with powers, either uniformly around it or arranged as the resultant of the powers of the two constituent atoms. . . .'

Faraday illustrated this in terms of two water waves impinging upon each other, combining, and then separating again.

Faraday may have derived this idea of chemical combination from his reading in *Naturphilosophie*. For we find rather similar ideas in Coleridge's work. Coleridge, claiming the authority of Heraclitus and Bruno, wrote:

> 'the Identity of Thesis and Antithesis is the substance of all Being; their Opposition the condition of all Existence, or Being manifested; and every Thing or Phenomenon is the Exponent of a Synthesis as long as the opposite energies are retained in that Synthesis. Thus Water is neither Oxygen nor Hydrogen, nor yet is it a commixture of both; but the Synthesis or Indifference of the two: and as long as the copula endures, by which it becomes Water, or rather which alone is Water, it is not less a simple Body than either of the imaginary Elements, improperly called its Ingredients or Components. It is the object of the mechanical atomistic Philosophy to confound Synthesis with synartesis, or rather with mere juxtaposition of Corpuscles, separated by invisible Interspaces. I find it difficult to determine, whether this theory contradicts the Reason or the Senses most; for it is alike inconceivable and unimaginable.'[40]

This notion reappeared in the Chemical Society's debates on atomism in 1867 and 1869; and Clerk Maxwell, in an essay on atoms, made the suggestion that there was no reason why two Boscovich atoms should not occupy the same space.

Whewell criticised the Boscovicheans because he thought that a world composed of point atoms would not be a real world, since such atoms could have no mass. In 1846, in a paper entitled 'Thoughts on Ray-vibrations', Faraday turned to the rebuttal of this charge.[41] Solid atoms or nuclei were no more required for copper atoms, which gravitate, than for ætherial ones, which do not:

'for of all the powers of matter gravitation is the one in which the force extends to the greatest possible distance from the supposed nucleus, being infinite in relation to the size of the latter, and reducing that nucleus to a mere centre of force'.

M. Koyré has recently made the point that a physics of central forces presupposes some kind of atoms; but Faraday is quite right in remarking that these atoms could be point atoms, and that gravitation is a good example of the mutual actions of centres of force. He added that we should not allow gravitation and ponderability to confuse us, for these phenomena arise not from the nuclei but from the forces. Gravitation and solidity are not due to the weight and contact of ultimate particles, but are the consequences of attractive and repulsive powers.

Faraday remained an atomist of a kind despite his trenchant criticisms, on both philosophical and experimental grounds, of the Daltonian position. Whewell, who like Faraday held the office of President of the Chemical Section of the British Association, and was for a time Professor of Mineralogy at Cambridge, went further in his refusal to permit atomic explanations. It was one of the most important of his objectives to link chemistry firmly to mineralogy; and hence he objected to Dalton's atomic diagrams because they lacked a basis in crystallography. Wollaston had also been interested in this last science, and as well as showing how the various crystal structures could be built up from spherical atoms he had invented an instrument, the goniometer, to measure the angles of crystals. Whewell followed him, in his *History of the Inductive Sciences* of 1837, in declaring that while Dalton's laws of chemical combination were nowhere contested, 'the view of matter as constituted of *atoms* . . . is neither so important nor so certain'. The term 'atom' he was prepared to retain for its convenience, provided no 'hypothesis of indivisible molecules'[42] was implied in its use. But certainly there was no room in the science for hypothetical configurations of atoms, which lacked all basis in the facts of chemistry or crystallography.

Whewell was much more taken with the electrochemical theories of Davy and Faraday than with the atomic theory, and remarked how workers had been drawn on, in their attempts to relate electrical polarity to chemical composition, by the promise of some deep and comprehensive insight into the mechanism of nature. And over the next few years his attitude seems to have hardened. A whole series of atomic theories were brought forward for censure in the *Philosophy of the Inductive Sciences*.

First comes the corpuscular philosophy, as presented by Dr Frend in a chemistry book published in 1710. Here Whewell's criticism is rather like Stahl's; the principles, he declared, were not applied in any definite manner to the explanation of phenomena. In other words, the treatment is general and superficial. Even more worthy of censure was the attempt to explain the complex properties of bodies by positing complex particles. This was begging the question, as Newton himself had said of hooked atoms as an explanation of cohesion; a mere disguising of our ignorance. The atoms, Whewell continued, were supposed to be held together by attraction, which was presumed to have some analogy with mechanical attraction. Yet:

> 'The doctrine that chemical "attraction" and mechanical attraction are forces of the same kind has never, so far as I am aware, been worked out into a system of chemical theory; nor even applied with any distinctness as an explanation of any particular chemical phenomena. Any such attempt, indeed, could only tend to bring more clearly into view the entire inadequacy of such a mode of explanation.

For the leading phenomena of chemistry are all of such a nature that no mechanical combination can serve to express them, without an immense accumulation of additional hypotheses. If we take as our problem the changes of colour, transparency, texture, taste, odour, produced by small changes in the ingredients, how can we expect to give a mechanical account of these, till we can give a mechanical account of colour, transparency, texture, taste, odour, themselves?'[43]

The atomists were trying to run before they could walk, and before the very possibility of walking had been established.

Pursuing his polemical course, Whewell would not even allow the atomists to triumph when the theory had been successfully applied to a problem. Laplace's calculations had shown that elasticity, capillary rise, and heat conduction could all be explained in terms of atoms with short-range forces between them. This was the sort of mathematical treatment which many of Whewell's contemporaries would have rejoiced in; but for Whewell it constituted no proof, nor even evidence, of the proposition that matter was composed of atoms. Coincidences between observation and calculations based upon a theoretical model could prove nothing. In particular, assumptions made in mathematical reasonings tell us nothing about the world; they are simply hypotheses which serve to help reduce the phenomena to calculation. In working out the centres of gravity of bodies, after all, one employs the assumption that they are continuous and can be divided into infinitesimably small parts. Neither the hypothesis that bodies are continuous, nor that they are atomic, implies any physical reality; both are adopted simply for their convenience.

This seems a very extreme position. Whewell would not allow that the success of a mathematical theory involving unobservable entities is in any way evidence for the real existence of those entities. It would seem that he is guilty of a high redefinition of 'evidence', for no better evidence could be forthcoming for unobservables. This argument would therefore exclude all unobservable entities from the sciences, except insofar as they play a formal role in the mathematics. The sciences become systems of laws connecting observables, as positivists would like them to be. Whewell did not in fact go quite as far as this, in that he admitted that the success of a theory does give us a strong assurance that its essential principles are true. When at the end of the nineteenth century the atomic theory proved successful over a very wide front, most people sensibly allowed themselves to be satisfied with a strong assurance that atoms existed, and did not worry that the evidence necessarily fell short of proof.

Maxwell in the introduction to his paper of 1866 on the dynamical theory of gases sought to refute this argument of Whewell's on the grounds that the two cases selected by Whewell were not comparable. Only the particulate theories had been proposed as candidates for reality:

> 'In certain applications of mathematics to physical questions, it is convenient to suppose bodies homogeneous in order to make the quantity of matter in each differential element a function of the co-ordinates, but I am not aware that any theory of this kind has been proposed to account for the different properties of bodies. Indeed the properties of a body supposed to be a uniform *plenum* may be affirmed dogmatically, but cannot be explained mathematically.'

Maxwell therefore thought that the successes of his theory were evidence for the existence of molecules; though he was careful not to assert the existence of unsplittable atoms.

The decision, in Whewell's view, of the question whether the world is composed of atoms or not must depend not upon chemical facts or mathematical deductions but on our notion of substance. It is a metaphysical question; still the same one which faced the Greeks, with all the uncertainties in its solution of science done in an armchair rather than a laboratory. Dalton's atomic theory gave rise to all sorts of problems, such as the half-atoms which appear in certain reactions, and crystallographic difficulties. The half-atoms problem disappeared if equivalents were used; this argument was earlier used against atomism by Berzelius in 1813. Dalton's hypothetical diagrams did not square with the evidence provided by the study of crystalline forms. Were the structure of calcium carbonate as Dalton depicted it, for example, then a crystal of it would be an 'oblique rhombic prism or pyramid' in shape; whereas its crystals are found to have a triangular symmetry, and to belong to the rhombohedral system. Whewell hoped that the progress of crystallography would parallel that of the understanding of the laws of chemical combination:

> 'Hypothetical arrangements of atoms, thus expressing both the chemical and crystalline symmetry which we know to belong to the substance, would be valuable steps in analytical science; and when they had been duly verified, the hypotheses might easily be divested of their atomic character.'[44]

Atomic hypotheses had an heuristic value, in that they might lead to discoveries; but then they should be dropped, and not permitted to become a part of the fabric of the science.

Whewell was as sweeping as Faraday in dealing with the notion that the hardness of bodies might be due to that of their particles; the nature of this atomic hardness being left unelucidated. 'What progress do we make,' he thundered, 'in explaining the properties of bodies, when we assume the same properties in our explanation?'[45] Faraday and Davy had escaped from this dilemma by embracing the Boscovich atom; but Whewell would not allow this model to be a candidate for reality. A collection of centres of force could not, in his view, possess inertia. Matter 'implies not only force but something which resists the action of force'. On the other hand, Boscovich's theory was homogeneous and consistent, and might be used as an instrument of discovery. Unfortunately it appeared to embody the deplorable confusion which Whewell had elsewhere detected, between chemical 'attraction', a theoretical concept, and mechanical attraction, an observed phenomenon.

Whewell, then, was prepared to accept atomic theories only as fictions. If natural philosophers found that to think in terms of atoms helped them to make discoveries or to organise diverse facts into a coherent picture, well and good; but they should then seek, in the actual presentation of their discoveries or laws, to remove such hypothetical matter. The possibility of atomism could be ruled out on logical grounds; it could only be an hypothesis in the sense in which Koyré expounds the use of the word in Newton's 'hypotheses non fingo'; the positing of falsity in order to deduce truth.[46] The assumption of atoms was bound to lead, in Whewell's view, to the pushing back of mysteries, and not to any solution of them; or else to ideal worlds much simpler than the real world, and easier to handle mathematically. No evidence could establish that atoms existed; definite proportions did not require an atomic explanation, nor did the observations of Faraday and Wollaston on evaporation and the extent of the Earth's atmosphere. And finally no calculations in which the postulation of atoms led to the prediction or explanation of phenomena cast any light on the actual structure of the world. Such mathematical artifices could at best save the appearances; and a number of hypotheses might be found which would do this. Indeed it might be convenient in some part of a science to presuppose atoms, and in another a continuum. The success of an assumption could only give psychological assurance of its likelihood; never could it supply real evidence. Whewell was in a long tradition, going back to Greek views on mathematical astronomy, in asserting that any mathematical hypotheses could be used in calculations; but that to know what was the real state of affairs required metaphysical investigation.

Sydney Smith remarked that while Whewell's forte was science, his foible was omniscience; and apparently his books were only moderately well received by scientists. In the next chapter, though, we shall follow physicists further in their efforts to achieve atomic theories which were not open to the kind of objections which Davy, Faraday, and Whewell had levelled against Daltonian atomism. In particular, atomic models were sought which, having very simple properties themselves, could explain the varied physical properties of things by their different arrangements alone; the classic programme of atomism in physics.

4

Some Theories of Matter

IN THIS chapter we shall be first concerned with theories designed to explain the properties of bodies, starting from first principles. These atomic theories fall into two groups; those such as Mosotti's involving more than one kind of fundamental particle, and the theory of the vortex atom, which was a compromise between an atomic theory and the view that matter is a continuum. These theories were developed by physicists and although their authors sought in a general way to account for the chemical properties of bodies, chemists seem to have paid little attention to them. Such theories solved no chemical problems, and therefore belong on the physical side of the divide between the two sciences; the object of all their authors was to postulate particles as simple as possible, which would nevertheless explain a wide range of properties. Also in the realm of physics was the dynamical or kinetic theory of matter, the object of which was to account in detail for the properties of gases—and in general for those of solids and liquids—on the assumption that they are composed of particles in rapid motion. Those responsible for this theory tended, like the chemists, to be reluctant to commit themselves to an *atomic* theory, and rested content with having made it very probable that matter was made up of molecules or particles. The theory is of great importance in our story because its conclusions were in accord with those deduced from the facts of chemistry; and we shall therefore pass on to it after discussing the vortex atom.

The cantankerous mathematician Charles Babbage—who was, like Whewell, suspected of omniscience—included in an appendix to his *Ninth Bridgewater Treatise* a discussion on atomism. The Bridgewater Treatises were a series of works, which appeared in the 1830s, on natural theology, in which the existence of God and His wisdom and benevolence were deduced, or at least argued for at length, from the order which exists in the world. Eight authors were commissioned to write these books, and received substantial payment for doing so under the will of an Earl of Bridgewater; but Babbage considered that mathematical physicists had been inadequately represented in the list, and so wrote a ninth unauthorised treatise. In many ways it is the most interesting of the series.

In the text of the work, Babbage wrote that

> 'all analogy leads us to infer, and new discoveries continually direct our expectation to the idea, that the most extensive laws to which we have hitherto attained, converge to some few simple and general principles, by which the whole of the material universe is sustained, and from which its infinitely varied phenomena emerge as the necessary consequences.'[47]

The development of the sciences of astronomy and optics made this seem likely. The appendix elaborated this notion.[48] 'Ever since Newton's day', wrote Babbage,

> 'natural philosophers had hoped that some law even more comprehensive than his law of gravity might be discovered; a law of which gravitation, among other phenomena, would be a consequence.'

By the 1830s it seemed that what for Newton had been a hope might soon become a reality, in Mosotti's atomic theory; described by Babbage as not a 'happy conjecture' but an hypothesis 'carried through the aid of profound mathematical reasoning, to many of its remote consequences'.

Mosotti[49] had supposed the world to be made up of two kinds of particles or centres of force; atoms of matter and atoms of ether, each of which repel similar particles and attract dissimilar ones. The attractive and repulsive forces both obey the inverse square law. By making the repulsive force between atoms of matter slightly less than the attractive force between the atoms of ether and those of matter, Mosotti was able to derive an inverse square law of attraction between bodies corresponding to the law of gravity. The phenomena of electricity could also be accounted for on his assumptions. Faraday was enthusiastic about this simplification, which seemed a step towards the fulfilment of the programme which Davy had laid down in his lectures and Faraday supported. An atomic theory which involved the minimum of assumptions, very few different kinds of atom, and the promise of mathematical treatment, should have had a wide appeal. Faraday believed that Mosotti's scheme 'fell in very harmoniously' with his own Boscovichean views.

Babbage believed that it might be of interest to investigate what should be expected of this general law to which all phenomena could be referred. If a stream of ether particles were directed at a matter particle, or central atom, surrounded with an atmosphere of ether, undulations would be produced throughout the whole atmosphere, and the whole outer layer, pressed 'beyond the limit of attraction', would fly off at right angles. The atmosphere could also transmit vibrations in all directions in space filled with etherial particles. In one of these ways radiation could be explained. When several particles of matter, each with its atmosphere, are considered, then the

SOME THEORIES OF MATTER

theory must account for the various states of matter, and the properties of these states; for example, the cohesion and elasticity of solids, and the capillarity of fluids. The particles will be attracted towards each other by a force depending on the difference between the attractions of the particles for each other's atmospheres, and the repulsion between the particles themselves.

In order to explain chemical union, the theory must allow for the matter particles to approach more intimately, so that their atmospheres coalesce. Such compounds, composed of two or more atoms, would have their atmospheres not quite spherical, and would therefore be polar. And also, presumably, liable to decomposition by the electric current, although Babbage does not say so. The equations of the surfaces of the atmospheres of the various compounds might, Babbage hoped, 'perhaps, become the mathematical expression of the substance it constituted'. The chemical properties of bodies might then be deduced from the form of their characteristic surfaces; and it might be found that the heat given off in a reaction would depend upon the reduction in the amount of atmosphere which the atoms could hold when they combined, compared with that when they were separate.

Babbage's hope was that:

> 'Hence the whole of chemistry, and with it crystallography, would become a branch of mathematical analysis, which, like astronomy, taking its constants from observation, would enable us to predict the character of any new compound, and possibly indicate the source from which its formation might be anticipated.'

This is once again the hope of Davy, and of Babbage's friend Herschel. In these passages Babbage, like Boyle before him, does not seem to be quite clear how one gets from the identical atoms or centres of force to the chemical atoms which appear to be under discussion here.

Babbage declared that although in his exposition of Mosotti's theory he had, for the sake of simplicity, employed but two kinds of atom, he believed that three kinds might turn out to be necessary. Central atoms repel each other, and attract the other two kinds, which either repel, or are perhaps indifferent towards, each other. Each central atom would then have two layers of atmosphere around it.

> 'Under such circumstances, the outer atmosphere might give rise to heat and light, to solidity and fluidity, and the gaseous condition; to capillarity, to elasticity, tenacity, and malleability. The more intimate union of the central atoms, by which two or more become enclosed in one common atmosphere of the second kind, might represent chemical combinations, and perhaps that atmosphere itself be electricity. Possibly, also, this intermediate atmosphere, acted on by the pressure of the external one, and by the attraction of the central atom, might take the liquid form. These binary or multiple-combinations of the original atoms, and their smaller atmospheres, would still be enclosed in an atmosphere of the outer kind, which might be nearly spherical. The joint action of the three might, at sensible distances, produce gravity.'

Babbage merely threw out these suggestions as hints, for nothing but profound mathematical researches could, in his opinion, establish them or even give them a temporary value as an hypothesis covering a wide range. And it is certainly true that the theory seems more like a presentation of a suggestive model which might be worth exploring, and might have conceptual advantages over its rival theories of matter, than like a theory which could be in any sense experimentally falsified.

The notion of two kinds of particle, repelling similar and attracting dissimilar particles, was not original with Mosotti. It is to be found in Henry Cavendish's paper in the *Philosophical Transactions* of 1771, 'An Attempt to explain some of the Principal Phenomena of Electricity, by means of an Elastic Fluid'. Cavendish, taking as an hypothesis the existence of these two kinds of particle, some of matter and others of electric fluid, was able to deduce in an elegant manner various consequences which were confirmed by electrical experiments. What was new in Mosotti's work was the attempt to extend this kind of model to explain gravitation and the radiation of light as well as electricity; and also to incorporate the notion that the atoms might be mere centres of force, making the speculations acceptable to Faraday. Again, Cavendish had not committed himself on the question of the real existence of the elastic fluid of electricity, but had used the hypothesis as a convenient artifice only; whereas the theory of Mosotti and Babbage seems to have been seriously intended as a description of the world.

Once three kinds of particle are admitted, the theory begins to lose in conceptual simplicity, and compares rather badly with the scheme set out by Davy in his *Consolations* dialogue. On the other hand, the Boscovicheans could not supply any quantitative support for their theory, despite Davy's hope that adoption of the centre-of-force atom would make chemistry a mathematical science; whereas the theory of Mosotti did promise to be quantifiable in some areas at least. The chemical explanations sketched out by Babbage are, to say the least, vague, and at best hints of the form an explanation might one day take. But the hint was taken up, and in a series of papers Thomas Exley sought to apply a theory of this kind to the explanation of chemical and physical phenomena. Unfortunately, in his attempts to account for a wide range of chemical and physical properties, he found that he had to depart further and further from the pristine simplicity of Boscovich or of Mosotti, and to postulate more and more distinct kinds of particle. It is easy to sketch simple schemes; but complications come in when such schemes have to be made to fit the details.

Exley was a clergyman who published theological works as well as his papers on the atomic theory.[50] He wrote in 1837 that he was nearly seventy years old, and that if he were only but thirty he would follow up the theory that he could then only describe in broad outline. Nevertheless he was spared long enough to read a paper on the subject to the British Association in 1848. His first publication on the matter, a book on natural philosophy, appeared in 1829, before Mosotti's theory; and in this work he declared that the 'principle of attraction' pervaded the natural world, and was found wherever there was matter. Examples of attraction included chemical affinity, electricity, and magnetism. Repulsion was equally universal; so both attraction and repulsion must be supposed, according to Newton's Rules of Reasoning in Philosophy, primary properties of matter. Every attraction can be reduced to gravitation; and the repulsive force also probably follows an inverse square law. Solid atoms were not necessary, for repulsive forces rising to an infinite value, like those of Boscovich, would produce the effects of solidity. According to Exley, his theory resembles that of Boscovich: 'but the intermediate alternate spheres of attraction and repulsion are rejected'. Exley was in a rather more amateurish way than Faraday or Mosotti trying to achieve the same programme, a theory of matter which would start with very simple atoms, and would also indicate the common basis of the various attractive forces in nature. A world with both very simple basic atoms and very simple forces would be indeed a desideratum.

Gravitation, on Exley's view, was the thread which one should follow as guide through the sciences. The mechanical properties of matter he proposed to account for on the supposition that atoms differ in the radii of their spheres of repulsion and in the force at a given distance from their centres. When trying to explain, on this model, chemical reactions, Exley reads like a physicist dabbling in the subject, and manages to provide only vague, general, and *post hoc* speculations: this was, of course, in varying degrees the fate of all those who tried at this period to provide a chemistry based on first principles. The decomposition of potash by the electric current he described as follows: the electric fluid adheres strongly to the oxygen, which goes to the positive pole, while the separated metallic base is propelled towards the negative. Faraday wrote of Davy's explanation of this phenomenon that a dozen detailed incompatible schemes could be proposed, all consonant with Davy's suggestions and claiming to follow from them; but Davy's scheme is precision itself compared with Exley's.

But in his later papers, read to the British Association, Exley showed himself increasingly concerned with the phenomena of chemistry; and these essays were intended as steps towards a deductive chemical science. The atoms were supposed to have a repulsive sphere and an attractive one; and to differ both in the radii of their repulsive spheres and their absolute forces. Exley believed that in rejecting both the alternate spheres of attraction and repulsion of Boscovich, and the hard atoms of Newton, he had avoided launching into conjecture. By now Exley required two kinds of atom; a ponderable and tenacious species, composing ordinary matter, and an etherial kind, making up heat, light, and the electric fluid. As early as 1829 Exley had urged the inaccessibility of matter:

> 'We know nothing of matter, but by the forces which it exerts, and which doubtless constitute its nature.'

This is the sort of thing which Faraday said; but Faraday was perhaps more logical, therefore, in following Boscovich and making the particles of matter mere points.

Although his model appeared, to him at any rate, to have a firmer empirical base than those of Boscovich and Newton, without some additional hypothesis it was not possible to deploy it in chemistry. Exley realised this, and in his latest paper, in 1848, asserted that

> 'it has been of late too much the fashion to discard hypothesis. Newton discovered the law of gravitation; but how? by first admitting it hypothetically, and then testing the hypothesis by calculation'.

Dalton, it will be recalled, also invoked the example of Newton in his apology for his hypothetico-deductive mode of proceeding.

Exley had been using distinctly hypothetical explanations before 1848; for example to account in 1838 for the recently discovered phenomenon of catalysis. Hydrogen and oxygen combine without ignition in the presence of platinum. Platinum is dense and metallic, and therefore its atoms have small spheres of repulsion, and hence dense etherial atmospheres. These atmospheres prevent the oxygen atoms from combining with the platinum atoms, to which they approach as close as they can. The hydrogen atoms are also drawn towards the platinum; and combine there with the oxygen. This explanation, one need hardly add, is *ad hoc*, and completely fails to explain the unique efficacy of platinum. But all contemporary theories of catalysis were open to these objections.

In 1844 Exley had found it necessary to increase the number of his species of atom to four; tenacious (ordinary matter), electric (less force, but larger repulsive sphere), etherial (still less force, and even larger sphere of repulsion), and microgenal (very small force and repulsive sphere). The purpose of these last is rather obscure. The non-tenacious atoms, as in Babbage's schema, are arranged in shells beyond the sphere of repulsion of the tenacious atoms, and produce the alternate attractions and repulsions of the Boscovich atom. The latter, although admittedly these spheres are not explained, by now begins to appear considerably more economical; but Exley's attempt to link the Boscovich atom to a more particulate theory is interesting. He claimed that he could deduce that the pressure of a gas, whether elementary or compound, is proportional to the number of atoms present. This appears to be Avogadro's law, if by 'atom' here Exley meant what we mean by 'molecule'. Exley drew up rules for gaseous combination, which he described as his 'phenomena'. Either two simple or compound atoms combine, and the volume is unchanged or sometimes halved; or two atoms combine with intervention of a third, in which case the volume after reaction is the same as, or half, that of the reacting gases.

All this illustrates the muddle which resulted from the application, by a somewhat unsophisticated theorist, of atomic theories like those of Mosotti and Babbage in chemistry. Exley's theory seems more removed from the world of the chemical laboratory than was Dalton's, and positivists like Wollaston would have been able to make the same kind of criticisms against it. But whereas Dalton seems to have been content to allow the chemical elements to be all irreducibly different, Exley achieved a simpler conceptual scheme, in which the atoms of the different elements are distinguished first in terms of differences in forces and in spheres of repulsion, and then these are explained by introducing the four different kinds of atom, which can be grouped in a variety of ways. Admittedly, the applications to actual chemical phenomena are vague and indefinite; Exley lacked the grasp of the problems which professional workers must have, though few of them in fact did any better.

Nobody seems to have taken much notice of Exley's theory, but a few years later came another attempt to solve the scientific and philosophical problems of atomism which did arouse a great deal of attention. William Thomson—later Lord Kelvin—proposed that atoms were vortices in a perfect etherial fluid; and this notion was later described as 'by far the most fruitful in consequences of all the suggestions that have hitherto been made as to the ultimate nature of matter'.[51] Thomson oscillated between various kinds of atomic theory during his life; early on he had been a Boscovichean; but in 1884 he remarked that 'we have long passed away from the stage in which Father Boscovich is accepted as being the originator of a correct representation of the ultimate nature of matter and force'. Then twenty years later he added a note to the effect that the theory had been 'wrongly judged obsolete'.[52] His vortex atom proposals came in the period when he found Boscovich's model unsatisfactory.

In 1858 the German polymath Helmholtz wrote a mathematical paper on the motion of vortex rings in a fluid, which Peter Guthrie Tait later translated into English. Tait, who collaborated with Thomson to produce the first textbook of mechanics to be firmly based on the concept of energy, built a machine to make smoke rings; in fact a similar apparatus had been described by Babbage. One side of a wooden box was cut away, and replaced by cloth; and on the opposite side a round hole was cut. When the bottom of the box was sprinkled with a strong solution of ammonia, and a dish of salt and sulphuric acid put in, and the cloth struck smartly, smoke rings were ejected through the hole. It is possible, though rather difficult, to make small rings pass through the middle of larger ones; and to make the rings collide, when they bounce off each other elastically.

Confronted with Helmholtz's paper and Tait's machine, Thomson at once jumped to the conclusion that atoms were vortex rings in a perfect frictionless fluid medium. Merz, in his *History of European Thought in the Nineteenth Century*, wrote that he believed this

> 'the most advanced conception, in this line of thought, of which the human mind has so far been capable . . . one of the most remarkable instances of the fruitful reaction of abstract mathematical reasoning on the progress of physical research . . .'[53]

He noted that the problem was how to explain weight and inertia in terms of the vortices; the same difficulty which faced the adherents of the Boscovich atom. But only a few years later, in 1908, Thomson's obituarist remarked that nobody then believed that atoms were really vortex rings, and that there was some doubt whether Thomson ever thought they were. Nevertheless, his influence had been such that many others had followed him, and the theory had become a beacon-light for the newer developments in physics.

Thomson was severe towards atomists among the chemists.[54] The only evidence that they supplied for their hard atoms, he wrote, in 'their rashly worded introductory statements', was the argument of Lucretius and Newton that the properties of matter do not change. This was not convincing, for vortices in a perfect, frictionless fluid would be permanent too; and moreover: 'Lucretius' atom does not explain the properties of matter without attributing them to the atom itself'. Vortices, on the other hand, were elastic and permanent without having these properties specially ascribed to them; Maxwell thought it a great attraction of the vortex atom that it was impossible to add new properties to it *ad hoc* to save the appearances. Thomson amplified this criticism of orthodox atomism, which resembled that of Whewell, in his Presidential Address to the British Association in 1871.[55]

In that address he welcomed the kinetic theory of gases, but added a caveat:

> '... there can be no permanent satisfaction to the mind in explaining heat, light, elasticity, diffusion, electricity and magnetism, in gases, liquids, and solids, and describing precisely the relations of these different states of matter to one another by statistics of great numbers of atoms, when the properties of the atom itself are simply assumed ... what is the inner nature of the atom?'

One is reminded of his *cri de cœur* on another occasion: 'I want to understand light as well as I can without introducing things that we understand even less of'. A very reasonable philosophical position; it would be strange indeed to know less about one's model than about the phenomenon it is invoked to explain. Thomson thus criticised the explanation of the elasticity of gases in terms of elastic particles, for the elasticity of solids seemed a more complex matter, and was less understood, than that of gases.

Thomson believed that some heterogeneity was a feature of the world, and he calculated from four different pieces of evidence the limits of the sizes of atoms—or, in fact, molecules. This paper gave the order of size of molecules, and prevented chemists from assuming excessively small atoms, though it did not necessarily prove, as Thomson himself realised, that below the molecular level matter might not be infinitely divisible.

Thomson's flirtation with the vortex atom did not last very long. By 1881 he feared that the theory could not explain all the facts of chemistry; by 1884 he had become distinctly disenchanted with it; and by 1904 he was afraid that the vortices might not even be stable indefinitely. In 1884 he designed an arrangement of rigid rods and flywheels which behaved as a spring, in an effort to explain the elasticity of gases without postulating more complex elasticities. This device gave him an atomic model which explained elasticity in terms of rigidity and motion alone; but unfortunately rigidity was still left unexplained, and refused to be reduced to motion. The American positivist philosopher Stallo was as ruthless with vortex atoms as with any others; he thought them a splendid reduction of the known and familiar to the unknown and unknowable. To explain the familiar fact of action at a distance, we are offered a hypothetical medium with properties different from those of ordinary matter. An anonymous reviewer in the *North British Review* for 1868 was kinder, and saw in the vortex atom a reconciliation of the apparently opposed atomic and continuum theories.

The short history of the vortex atom was marked by an incursion of the young J. J. Thomson, who tried to apply the theory in an extremely interesting manner in chemistry. He explained the dissociation of gases, for example, in terms of paired rings whose radii were changed by disturbing influences so that they separated. And his view of the chemical elements deserves quotation:

> 'Now let us suppose that the atoms of the different elements are made up of vortex rings all of the same strength, but that some of these elements consist of only one of these rings, others of two of the rings linked together, or else of a continuous curve with two loops, others of three and so on; but our investigation . . . shews that no element can consist of more than six of the rings if they are arranged in the symmetrical way there described.'[56]

Valency—the combining ratios for different elements—can be correlated, on this view, with number of rings. A 'two-ring' element can combine with two 'one-ring' elements, for example, and the maximum value of six rings explains why the maximum valency observed is six. Even variable valency could be given some kind of explanation. And finally the theory could explain why it is that monovalent elements have the simplest spectra; they have the simplest structures. In the hands of J. J. Thomson, then, the vortex atom reached the high point of its career as a candidate for truth. Compared with Exley's vague theorising and Davy's Boscovichean dialogue, this account has the enormous advantage of being quantitative; and at the same time making very simple basic assumptions, like these earlier workers' theories had. It is sad that the vortex atom turned out to be a blind alley rather than the splendid highway which it had at first appeared to be.

Progress in fact came through the kinetic theory of gases,[51] and it was along this road that physics finally came to rejoin chemistry in the 1870s and 1880s. The earliest writings on the kinetic theory seem to have been those of John Herapath, published in 1821. Herapath believed in atoms; he wrote that

'the old notions of the divisibility of matter *ad infinitum* have been entirely dissipated under the light which chemistry has thrown on the subject'.

He even believed that his contemporaries were acquainted with bodies composed of simple atoms. The atoms must be absolutely hard; this seemed almost self-evident. Their shape could not be altered by impact or compression. And the forces which held them together were modifications of 'attraction': 'gravitational attraction being the attraction of distance, and chemical affinity that of contact'. Herapath's objective was to find, in a kinetic theory based on hard atoms, the simple and exceedingly comprehensive cause of origin of all the great agents of nature; including heat, electricity, and magnetism.

Herapath shared the conviction of Davy and others that matter must be simple. He wrote:

> 'One of the sublimest ideas of the ancients was, that there is but one kind of matter, from the different sizes, figures and arrangements of whose primitive particles, arises all that beautiful variety of colour, hardness and softness, solidity, and fluidity, opacity and transparency, &c. which is observed in the productions of nature. Our first two postulata [that atoms are hard, and compose solids and liquids according to their various associations] do not require that there should be but one kind of matter; there may be several kinds. But since it seems possible to account for all the phenomena on the supposition of one kind only, and since nature is always disposed to employ the simplest machinery, probability is strongly in favour of the ancient idea.'

The problem of the collision of absolutely hard, unyielding bodies defies the imagination, and Herapath, writing before the principle of conservation of energy had been announced, found himself in difficulties. Hard bodies striking hard planes, he concluded, would come to rest; but two hard balls, meeting with equal and opposite momenta, would separate with the same equal and opposite velocities. If one ball hit another stationary ball, the latter would move off with the same velocity as the former before the collision. Herapath was able, on this theory, to account for some of the properties of gases; but he was relatively unknown, his paper contained no prediction of any crucial experiment, and appeared speculative; it was therefore rejected by the Royal Society, and appeared in a less exalted journal where it seems to have aroused little interest. Davy, the President of the Royal Society, would have sympathised with the speculation on the unity of matter; but the notion of hard atoms would have been repugnant to one who adhered to the theory of Boscovich.

The kinetic theory was next taken up by Waterston, a billiards player, conscious of the fact that he was discussing an idealised model which might fit the real world. The theory applies, he wrote:

> 'to a medium composed of free and perfectly elastic molecules in a state of motion [which] we have carefully to refrain from assimilating to any known form of matter until, by synthetical reasoning, circumstantial evidence has been accumulated sufficient to prove or render probable its identity'.

His atoms, unlike those of Herapath, were elastic; and so were those of Maxwell and Clausius, scientists well-established enough for their theories to be bound to attract attention. Maxwell introduced statistical methods, and also made the prediction that the viscosity of gases would be found to be independent of their density; a prediction which was soon verified experimentally. All these later theories would seem to be molecular, rather than atomic, for elasticity presupposes parts, unless the atoms are mere centres of force, or perhaps vortices. Only Herapath's hard particles could have been real atoms in the classical sense. Clerk Maxwell in one paper tried to treat the particles as mere points with forces between them, in the Boscovichean tradition; but in general it seems fair to say that the particles of the kinetic theory were elastic corpuscles, open to William Thomson's criticism that the elasticity of gases was being explained only by the invocation of more complex elasticities. 'Elastic atoms' only became a reasonable proposition when the 'atom' of the chemists was shown to be a complex body, at the end of the nineteenth century; and 'atoms' and 'molecules' were not clearly and generally distinguished until after Cannizzaro's paper to the Carlsruhe Conference in 1860. Except in his famous article on atoms, Maxwell used 'molecule' wherever possible, defining the term as the smallest possible portion of a particular substance, while atoms were particles which could not be cut in two. In his paper read to the British Association in 1859 he began with a discussion of the status of the particles (ideal, small, hard, elastic spheres) assumed in the theory. He said:

> 'So many of the properties of matter, especially when in the gaseous form, can be deduced from the hypothesis that their minute parts are in rapid motion, the velocity increasing with the temperature, that the precise nature of this motion becomes a subject of rational curiosity.'

Maxwell was emphatic that in the kinetic theory a model was being compared with the world; but that if the properties of the congeries of spheres were found to correspond with those of gases, then 'an important physical analogy' would have been established, which might lead to more accurate knowledge of the properties of matter.

He refused to commit himself completely to the physical reality of the model, however, for in the article on atoms he pointed out that it was unfair to make the small, hard billiard-ball atom elastic, so that it could explain more phenomena; for elasticity was the very property which it had been originally devised to explain. The small, hard, elastic spheres of his paper of 1859 would not therefore do; and in his paper of 1866 he proposed complex molecules. In the earlier paper he had remarked that a model involving centres of force would have been equivalent to that employing small spheres:

> 'Instead of saying that the particles are hard, spherical, and elastic, we may if we please say that the particles are centres

of force, of which the action is insensible except at a certain small distance, when it suddenly appears as a repulsive force of very great intensity. It is evident that either assumption will lead to the same results.'

In Maxwell, as in Waterston, therefore, we find (as we did in Dalton) that it is by no means always clear whether he is really propounding a theory of matter; or perhaps it would be fairer to say that Maxwell was very conscious of the line separating fact from theory, and was not prepared to step across it. To say that matter was probably particulate was as far as he would go.

William Thomson, writing in 1867, hoped that the vortex atom might have all the positive properties required by the kinetic theory

> 'without requiring any other property in the matter whose motion composes them than inertia and incompressible occupation of space. A full mathematical investigation of the mutual action between two vortex rings of any given magnitudes and velocities passing one another in any two lines, so directed that they never come nearer one another than a large multiple of the diameter of either, is a perfectly solvable mathematical problem, and the novelty of the circumstances contemplated presents difficulties of an exciting character. Its solution will become the foundation of the proposed new kinetic theory of gases.'

Three years later, Maxwell wrote that in the vortex atom there was nothing arbitrary, no central forces or occult properties; nothing but matter and motion:

> 'when the vortex is once started its properties are all determined from the original impetus, and no further assump-

tions are possible. Even in the present undeveloped state of the theory, the contemplation of the individuality and indestructibility of the ring-vortex in a perfect fluid cannot fail to disturb the commonly received opinion that the molecule, in order to be permanent, must be a very hard body'.

Maxwell rejected hard atoms because they could not vibrate as we see from their spectra that atoms do; and he also believed that to explain spectra in terms of Boscovich atoms would be bound to be 'in questionable scientific taste'. Vortices seemed to avoid question-begging assumptions on both sides, and the most serious problem of the model was whether bodies composed of such atoms would have momentum and energy or not. But as we saw, the vortex atom did not justify the high hopes expressed at its birth; and the problem of the elasticity of atoms was solved otherwise.

The kinetic theory of gases slowly became accepted, though William Thomson was dubious about Maxwell's assumptions concerning equipartition of energy between the translational, rotational, and vibrational modes of motion of the molecules. Maxwell pointed out that the consistency between the chemists' notion of a molecule, and the dynamical one, was evidence for the real existence of such entities, for the two chains of evidence and reasoning were independent. In the 1890s the kinetic theory was also used by Rayleigh and Ramsay in their dramatic work on argon, the first of the family of inert gas elements to be discovered. They deduced from the ratio of the specific heats of the new gas, at constant pressure and at constant volume, that it was monatomic; that is to say, its molecules contained but one atom. William Thomson remained dubious; but almost everybody else was convinced. By this time Maxwell's hope that the kinetic theory would re-unite chemistry and physics was being realised.

An atomic theory approximately contemporary with the vortex atom and with Maxwell's development of the kinetic theory, was the hypothesis of Thomas Graham, a chemist best known for his work on the rates of diffusion of gases. In an essay in the same journal as Faraday's Boscovichean paper, and with a very similar title—'Speculative Ideas respecting the Constitution of Matter'—Graham proposed a dynamical, molecular theory of the chemical elements:

> 'It is conceivable that the various kinds of matter, now recognised as different elementary substances, may possess one and the same ultimate or atomic molecule existing in different conditions of movement. The essential unity of matter is an hypothesis in harmony with the equal action of gravity upon all bodies.'[58]

We are to imagine but one kind of ponderable matter, composed of uniform atoms, always more or less in motion, which is retained from a primordial impulse. The more rapid their motion, the greater will be the volume which they occupy. Matter differs in density, for the motion of atoms is inalienable, and so, therefore, is the space which they fill. Light and heavy matter are not interconvertible: 'In short, matter of different densities forms different substances—different inconvertible elements as they have been considered.' In gases the atoms or molecules were believed to be of a greater complexity. In the same paper, Graham toyed with another hypothesis involving the luminiferous ether. The essay is very short, and must have been intended entirely as a number of suggestions of a rather general kind and not as a theory in which any of the details had been worked out.

Graham's biographer R. A. Smith noted that a consistent Daltonian atomist—and Graham he considered as strict an atomist as could be found—might become 'a thorough non-atomist when he breaks up in his mind the present elements'.[59] The theory, in Smith's opinion, derived ultimately from Davy, an author with whom Graham was certainly familiar, in his work on acids at least. Davy had written that heat was a vibratory motion of atoms

> 'and one may almost say that it was the object of Graham's life to find out what the motion of an atom was. Davy does not define his vibratory motion, which could scarcely be imagined in an unconfined space unless the body had parts; for fluids he gives a motion of rotation making thereby a definite boundary; but we do not see in his opinions what an atom, say of iron, would do when alone.'

Davy would in fact have believed that an atom of iron, and indeed probably atoms of all known bodies, had parts, so an isolated atom could have vibrated perfectly well. Smith wrote of Graham—as might have been said of Davy—that 'in all his work we find him steadily thinking on the ultimate composition of bodies'. What is unexplained in Graham's theory is how the atoms retain their original motion unchanged. It seems probable that his hypothesis resembles the theory of gases of Newton or Davy; the particles are not in rapid translational movement, always colliding and gaining or losing energy at each impact, but are simply vibrating with different intensities, and do not come into contact with one another. They are running on the spot, rather than rushing hither and thither as envisaged in the kinetic theory of gases.

We have now explored some atomic and molecular theories in physics, down to a period when they began to tally with molecular theories independently developed in chemistry. But we have left the progress, or lack of it, in atomic and molecular theories in the latter science far behind. In the next chapter, therefore, we shall return to chemistry, to the molecular theories of Avogadro and his successors, to the radical theory, and to the phenomena of isomerism and isomorphism. We shall try to understand why it was that Avogadro's hypothesis, the acceptance of which half a century after its author had proposed it revolutionised theoretical chemistry, proved unacceptable for so long; and why no evidence from chemistry for atomism seemed wholly satisfactory before the 1860s at the earliest.

5

Chemical Molecular Theories

THE solution to the problem of how to avoid being arbitrary in assigning formulae to compounds was soon solved.[60] In 1809 Joseph Louis Gay-Lussac published a paper showing that gases combined in simple ratios by volume. He concluded that 'it is only in the gaseous state that substances are in the same circumstances and obey regular laws'. He had come to the study of gases because in them the 'attraction of cohesion' was absent; one would therefore expect that bodies in this state would obey simple laws. His hope was the same as that of Laplace, Davy, and numerous others; that he could

> 'give a proof of an idea advanced by several very distinguished chemists—that we are perhaps not far removed from the time when we shall be able to submit the bulk of chemical phenomena to calculation'.

Gay-Lussac proceeded to an analysis of the theories of chemical composition of his predecessors and contemporaries: Proust, who believed that pure substances always combined in exact ratios; Berthollet, whose view it was that the ratios were not, in general, exact, but depended upon the quantities of the substances present; and Dalton, with his notion of combination atom to atom. Gay-Lussac, anxious for compromise, decided that Dalton's theory was not the whole story. His own experimental researches showed that gases react in surprisingly exact ratios by volume; both oxygen and hydrogen to form water, and ammonia and a range of acidic gases to form salts. He noted that

> 'in considering weights there is no simple and finite relation between the elements of any one compound; it is only when there is a second compound between the same elements that the new proportion of the element that has been added is a multiple of the first quantity. Gases, on the contrary, in whatever proportions they combine, always give rise to compounds whose elements by volume are multiples of each other'.

Further, when the products of reaction are gaseous, their volume also bears a simple ratio to the volumes of the reactants.

Gay-Lussac considered that the results of his investigations were very favourable to Dalton's theory, which he tried to reconcile with Berthollet's; after all, the paper appeared in a journal of which Berthollet was editor. It is therefore surprising that, far from being acceptable to Dalton, it aroused his indignation. He had, in his *New System of Chemistry*, toyed with a rather similar notion. He wrote that, in common with many people, he had had the

> 'confused idea . . . that a given volume of oxygenous gas contains just as many particles as the same volume of hydrogenous'.

Clearly gases could not combine both atom to atom, and in simple ratios by volume, unless equal volumes contained equal numbers of atoms, or at least numbers in simple ratios. Let us investigate why it was that Dalton abandoned the idea having once raised it.

His attack on Gay-Lussac, in an appendix to his book, took the form of an attempt to undermine the theory behind the paper, and also to impugn the accuracy of the actual experimental observations. Dalton was no great experimentalist, and the prospect of him upbraiding one of the best observers in Europe for his alleged inaccuracies has entertained or outraged many commentators. Dalton wrote, as we saw earlier, that Gay-Lussac's results must not be accepted until some *reason* could be found for them. He declared that if it could be established that equal volumes of gases contain equal numbers of atoms, or numbers in simple ratios, then Gay-Lussac's volumes or 'measures' would be equivalent to 'atoms'; but the latter term, not being confined to gases, was the more universal. The problem was that when gases combine the results are not what this hypothesis would predict. Equal volumes of nitrogen and oxygen would, on this view, contain equal numbers of atoms; when they combined to yield nitrous gas, the number of compound atoms formed must be less than, at most one half of, the total number of oxygen and nitrogen atoms before combination. The volume of nitrous gas should therefore not exceed one half the total volume of its component gases before the reaction took place. But the total volume is unchanged. Therefore equal volumes of gases do not contain equal numbers of atoms.

Because they were theoretically impossible, then, Dalton blithely rejected Gay-Lussac's analyses, referring to his observations as 'hypothesis'. He wrote:

> 'The truth is, I believe, that gases do not unite in equal or exact measures in any one instance; when they appear to to do so, it is owing to the inaccuracy of our experiments.'

The nearest approach to 'mathematical accuracy', he believed, was the oxygen/hydrogen ratio in water; but even this was not quite 2 : 1 but about 1·97 : 1. Dalton could not of course explain why the ratios should happen in so many cases to approach so very close to whole numbers.

Those who were distrustful of the atomic hypothesis naturally disagreed with Dalton's interpretation, and 'volumes' became a respectable alternative to 'equivalents'. It was an observed fact that gases reacted in simple ratios by volume, but a mere hypothesis that all bodies combined atom to atom. The problem of the formation of nitrous gas would disappear if the atom of this gas were formed of one half atom of oxygen and one half atom of nitrogen, instead of one atom of each constituent. But, as Berzelius remarked, a half atom is an absurdity. A half volume, on the other hand, is easy to picture. And when in 1813 Berzelius proposed the chemical notation still in use,[61] according to which the elements are represented by the initial letter of their Latin names, each letter stood not, as now, for an *atom* but for the weight of a *volume*. More exactly, the letters represented: 'the specific gravity of the substances in their gaseous state, that of oxygen being considered as unity', and not the atomic weights for which Dalton's hieroglyphic symbols had been devised.

Berzelius believed that all atoms would have to be the same size, in order to produce regular crystalline forms; and that all compounds would have to consist of not more than one atom of one of the constituents, and one or more of the other. A compound containing two atoms of one component and, say, three of the other, would be liable, according to him, to mechanical division. Compounds such as the oxides of iron now written as Fe_2O_3, and Fe_3O_4 were impossible, and must therefore contain fractions of atoms. It was better not to speak of atoms at all.

Dalton believed that the scandal of half atoms could be avoided by a judicious sliding of one's scale of atomic weights. The atoms he considered to be of different sizes; but their effective sizes were the radii of their atmospheres of heat; and these were always approximately spherical, and touched one another in gases as in solids and liquids. Berzelius in reply contrasted his experimental approach, that of the 'ordinary man', with Dalton's *a priori* methods, characteristic of the 'inventor'. In 1814, Thomas Thomson remarked that neither Davy nor Berzelius were atomists; but by the third decade of the century Berzelius had become a supporter of Dalton's theory.

We must return to Gay-Lussac's paper, for the essay which succeeded in reconciling the results of Gay-Lussac with the hypothesis of Dalton was published in 1811 by the Italian scientist Avogadro. He perceived the weakness of Dalton's position in denying Gay-Lussac's observations, and wrote:

> 'If we were to suppose that the number of molecules contained in a given volume were different for different gases, it would scarcely be possible to conceive that the law regulating the distance of molecules could give, in all cases, relations as simple as those which the facts just detailed compel us to acknowledge between the volumes and the number of molecules.'

Avogadro was right; even Dalton had to admit that gases reacted very nearly in simple ratios by volume, and this would not have been likely had the numbers of molecules, or atoms, in a given volume been very different.

The modern distinction between atoms and molecules was not made by Avogadro, who used the term 'molecule' throughout. Dalton, we should remember, always employed the term 'atom'; a difference in nomenclature which tended to characterise the English chemists. 'Atom' and 'molecule' seem to have been first distinguished in the modern manner in 1833 by Gaudin, a disciple of Ampère. Gaudin defined 'atom' as a little homogeneous spherical body, essentially indivisible, while a 'molecule' was an isolated group of atoms of whatever number and whatever nature. But Avogadro used the term 'elementary molecule' to mean 'atom', distinguishing this consistently from other kinds of molecule. His hypothesis—now usually dignified with the title of 'law'—was that equal volumes of all gases, under the same conditions, contain equal numbers of molecules. And as a corollary to this, that the molecules of elementary gases were usually composed of more than one atom. To divide a molecule consisting of two or more atoms on combination presented no problems; whereas to halve an atom was not possible.

The hypothesis not only made it possible readily to explain Gay-Lussac's results, but also provided a method of finding true molecular formulae without arbitrary simplicity rules. As Avogadro himself wrote,

> 'we have the means of determining very easily the relative masses of the molecules of substances obtainable in the gaseous state, and the relative number of molecules in compounds'.

Since equal volumes contain equal numbers of molecules, the weights of molecules will be in the same proportion as the densities of the gases they compose. And the formulae of compounds can easily be settled from the volumes of reactants that went to form them, provided the simplifying assumption is made that molecules do not contain more atoms than the minimum demanded to explain the facts. Thus when oxygen enters into combination to form nitrous gas or water, its molecule must be, according to the hypothesis, halved; it must therefore contain an even number of atoms. This number is, according to the simplifying assumption, two; for the molecule never seems to be divided into more than two parts. This assumption could not, before the development of the kinetic theory, be checked; but it is clearly a much less alarming affair than Dalton's simplicity rules, which demanded both judgements on which was the simplest compound between two elements, and also that this simplest be regarded as binary.

From the proportions established by Gay-Lussac, Avogadro showed that the formula for ammonia was what we would write as NH_3, and also produced the right formulae for the oxides of nitrogen. In the formation of water, the oxygen molecule is divided into two, as we saw; and Avogadro allowed that some molecules might be larger, and capable of further division. His conclusion was that

> 'Dalton, on arbitrary assumptions as to the most likely relative number of molecules in compounds, has endeavoured to fix ratios between the masses of the molecules of simple substances. Our hypothesis... puts us in a position to confirm or rectify his results from precise data and, above all, to assign the magnitude of compound molecules according to the volumes of the gaseous compounds, which depend partly upon the division of molecules entirely unsuspected by this physicist.'

As we shall see, Dalton had what seemed to him good reasons for not suspecting the division of molecules, or rather their formation in the first place.

It is well known that Avogadro's hypothesis did not play an important role in chemistry until the 1860s, after its resurrection at the Carlsruhe Conference by Cannizzaro. It is sometimes implied that for half a century the hypothesis was unknown. In fact, this was not the case at all. Notable chemists 'rediscovered' it at intervals, but there were always good reasons why it could not be accepted, as Cannizzaro himself explained. In some way or other, until the 1860s, the hypothesis ran counter to some theory felt to be crucial in the science, or seemed incompatible with some experimental data, or fell foul of the tidy-mindedness of some systematiser. Indeed it seems rare in the history of science for a new theory to be opposed simply on the grounds of newness, or ignored, especially if it be proposed by established scientists, as Avogadro's hypothesis was. There is almost always a good reason why a theory, which later seems a splendid anticipation of modern developments, was unsatisfactory to contemporaries; and in the case of Avogadro's hypothesis there was a whole array of good reasons why chemists did not seize upon it.

The hypothesis appeared, apparently independently, in a letter written three years later by the to physicist Ampère Berthollet, the dean of French chemists. The letter also contained some speculation on the shapes of molecules. Ampère put four atoms in a number of molecules where we would put but two; his simplifying assumption was less rigorous than that of Avogadro and Cannizzaro. Though Avogadro's paper was earlier and seems superior, Ampère's letter was published and became better known to later workers. Dumas used the hypothesis for a time, and so did Gaudin; while in England William Prout published it in his Bridgewater Treatise on *Chemistry Meteorology and the Function of Digestion*, claiming that he had discovered it independently.

Dalton did not accept Gay-Lussac's results, and therefore was not faced with the problem of reconciling them to his theory. Had he been more open-minded, he would have still found it difficult to accept molecules composed of two atoms of the same kind, for he believed that such atoms repelled each other. This notion originated in his thinking about the atmosphere; in particular, he found it hard to understand why the atmosphere was not a sandwich, with the heaviest gas at the bottom, and the lightest at the top. Research showed that its composition was uniform, and did not change with height above sea level; and Dalton concluded that this degree of mixing could only be brought about if atoms repelled like atoms, but not unlike ones.

Davy criticised this theory, on the grounds that, were it true, it would be impossible to keep hydrogen gas in an inverted beaker, since the repulsion between the hydrogen atoms would force those at the open end downwards and outwards. Dalton seems to have later modified this theory, and made it more mechanical. Differences in particle size as well as repulsive forces were invoked to explain the way the gases of the atmosphere became mixed up; but repulsive forces of some kind were retained. Prout appears to have taken over this notion, but he applied it to molecules rather than to atoms. His statement of Avogadro's hypothesis refers to 'self-repulsive molecules'. He deduced the hypothesis from physical evidence, the uniform expansion of gases with rising temperature, rather than from chemical reactions. Dalton, since his atoms repelled similar atoms, could not have easily accepted molecules of great stability composed of two or more such atoms.

The same was true of those who adhered to the electrical theory of chemical bonding which Davy had originated, though Davy himself was flexible enough to use something like Avogadro's hypothesis. If potassium, for example, combined with oxygen because it had a positive charge and oxygen a negative, then two atoms of the latter gas could not combine together because they would have like charges and like charges repel. In the hands of Berzelius, this notion became the system, dualism, which dominated chemistry for a generation. All chemical bonds, in the organic as in the inorganic branch of the science, were explained as due to the different electrical charges on the atoms.

With the rise of organic chemistry, this mechanism became

increasingly less likely; and when it was shown that in acetic acid electropositive hydrogen atoms could be replaced by electronegative chlorine ones without radical change of properties, dualism was doomed. Berzelius sought to escape from this difficulty by designating hydrocarbon groups as 'copulae', in which the ordinary laws of chemical bonding were not obeyed. This was attacked by critics as 'flagrant dishonesty', though it might be thought of as quite a sensible way of discriminating between the hydrocarbon part and the other groups in organic acids, bases, and so on. But in the 1840s evidence was provided for the existence of chemical bonds between like atoms, first by Silbermann, who showed that carbon gave off more heat when burned in nitric acid than in oxygen, and interpreted the difference as due to the heat required to rupture the O–O bond; and secondly by Brodie, whom we shall meet again. He provided arguments in support of links between like atoms on theoretical grounds, since elements did not behave very differently from compounds. Brodie believed that at the moment of chemical change this bond between like atoms became polarised, so that one atom became positive and the other negative. This would be easier to envisage, as he remarked, if the chemical atoms of the elements were not true atoms at all, but were molecules, made up of 'other and further elements'.[62] Complex chemical atoms could readily become polarised, whereas it is hard to imagine how this could happen with billiard-ball atoms having no parts. Brodie referred to Ampère's molecular theory as philosophical and suggestive, though still hypothetical.

So by 1850 dualism was no longer a major obstacle to the acceptance of the hypothesis of Avogadro or Ampère. Other difficulties had by now begun to make themselves felt. Dumas began as an enthusiastic supporter of the hypothesis. In 1826 he began his experiments on vapour densities; and in 1828 he declared that:

> 'Gases, under similar conditions, are composed of molecules or of atoms the same distance apart, that is to say that they contain the same number in the same volume.'[63]

He based this conclusion on both physical and chemical evidence; and saw the implication that

> 'each atom of chlorine, in combining with an atom of hydrogen, can only produce one atom of hydrogen chloride ... it must then be admitted that the atoms of chlorine and hydrogen are cut in two to form atoms of hydrochloric acid. Each of these last is composed then of a half-atom of chlorine and a half-atom of hydrogen.'

Although his terminology differs from ours, and seems indeed to be rather confusing in its use of the term 'half-atom', nevertheless Dumas seems in this passage to be saying what chemists since Cannizzaro have said.

By 1837 his attitude had changed, and he declared that the term 'atom' should be abolished because it went beyond experience, which chemistry should never do. By this time he had come to believe that the chemical elements were complex, and for this reason opposed the atomic theory; and his work on vapour densities had run into difficulties. Some elements, particularly sulphur and phosphorus, appeared to have different vapour densities, and therefore different atomic weights, at different temperatures.

This difficulty was not resolved for twenty years, until the work of Deville and others on thermal dissociation, the process by which compounds break up reversibly into simpler substances, or large molecules into smaller ones, at high temperatures. Sulphur and phosphorus vapours at relatively low temperatures are composed of large molecules; and as they are heated these dissociate into small ones, and the molecular weight, and hence vapour density, changes. The number of *atoms* in a given volume of the vapour therefore varies widely with temperature.

By 1860, then, these various experimental and theoretical objections, produced by those who had at least begun as atomists, to Avogadro's hypothesis had disappeared; but there still remained a residue of chemists who refused to support an atomic theory at all, and therefore could not accept a molecular theory either. The reasons for their opposition to atomism were the same, broadly speaking, as their predecessors' a generation earlier. Some were positivists, and therefore rejected all explanations in terms of unobservable entities, while others believed in the simplicity and harmony of nature, and the ultimate unity of matter. This latter group received encouragement from the development in organic chemistry of the 'radical theory', announced by Dumas and Liebig in 1837.[64]

These authors remarked on the progress of inorganic chemistry which had followed the acceptance of Lavoisier's definition of an element. But this progress did not depend upon the elements being in fact indecomposable bodies. If they were analysed, 'nothing would be changed in the architecture of the monument, although its foundations would be more profoundly excavated'. The problem was to apply a similar conception in organic chemistry, where most compounds are made up only of the elements carbon, hydrogen, oxygen, and nitrogen. Dumas and Liebig suggested that, in fact, organic chemistry had its elements just like the inorganic branch of the science, but that they were 'radicals', groups of atoms which remained together through whole series of chemical reactions, such as benzoyl, C_6H_5CO. The 'final elements', carbon, hydrogen, oxygen, and so on, only appeared when all trace of organic origin had disappeared. Their conclusion was magnificent in its simplicity:

'In mineral chemistry, the radicals are simple; in organic chemistry the radicals are compound; that is all the difference. The laws of combination and of reaction are otherwise the same in these two branches of chemistry.'

Soon this argument was turned on its head. Robert Kane, a noted Irish chemist, President of Queen's College, Cork, and an editor of the *Philosophical Magazine*, believed that the existence of radicals which behaved in every way like elements gave considerable probability to the notion that the elements of inorganic chemistry were in fact radicals too, composed of very few different kinds of genuine elements. The only 'philosophical' distinction between the simple and the compound radicals was that as yet chemists did not know the composition of the former class. Other authors, including Dumas himself, thought much the same as Kane; and as late as the 1880s a series of Presidents of the Chemical Section of the British Association raised the same questions in their addresses. Davy had earlier argued in the same way, years before the radical theory had been proposed; but in the 1830s and 1840s the time was especially ripe for these speculations because two phenomena had been discovered which seemed to require explanation in terms of arrangements of atoms or particles in space. These phenomena were isomerism and isomorphism; and it seemed not unreasonable to hope that their study would cast light on the whole question of the relationship between atomic constitution and chemical properties.

In 1824 Liebig and Wöhler analysed fulminic and cyanic acids respectively and Gay-Lussac found that the analyses were identical. And at much the same time Faraday discovered the hydrocarbon butylene, which gave the same analysis as ethylene. Berzelius in 1832 distinguished these observations, and explained them in terms of the atomic theory.[65] He remarked that

> 'it was long taken as axiomatic, that substances of similar composition, having the same constituents in the same proportions, necessarily must have the same chemical properties. The investigations of Faraday appear to indicate that there may be an exception to this if two similarly composed substances differ in that the composition of one contains twice as many elementary atoms as occur in the other, although the proportions between the elements remain the same.'

Berzelius then passed on to the discovery of Liebig and Wöhler, and wrote:

> 'Recent researches have now shown that the absolute as well as the relative numbers of elementary atoms may be the same, their combination taking place in such a dissimilar way that the properties of equally composed bodies may be different.'

No difference in composition between cyanic and fulminic acids had been discoverable, and their molecular weights were the same; and so Berzelius, believing that specific ideas should have definite names, described as 'isomeric' (composed of equal parts), 'substances of similar composition and dissimilar properties', and the phenomenon was called isomerism.

A little earlier Mitscherlich had formed the hypothesis that

> 'different elements united with an equal number of atoms of one or more other common elements . . . crystallise alike, and that the likeness in crystal form is determined entirely and completely by the number of atoms and not by the differences of the elements'.[66]

Experiment seemed to confirm this hypothesis completely, and it became the law of isomorphism; that substances similar in crystalline form have the same number of atoms arranged in the same way in their crystals. This law promised to provide some reasonable basis for going from equivalent weights to atomic weights without using Dalton's arbitrary rules, for the formula of one compound could be deduced if the formula of another whose crystals had the same form were known. Atomic weights could be compared directly if only one constituent were different in the two isomorphs. Mitscherlich also discovered that some substances crystallise in two different forms; this phenomenon he named dimorphism, and explained it in terms of two different arrangements of the same atoms.

These two developments marked the first extension of the atomic theory beyond the mere 'explanation' of definite proportions; but both of them were equivocal and disappointing in the support which they gave to the atomists. Until after Cannizzaro had successfully resurrected Avogadro's hypothesis, chemists did not agree on atomic weights, and it was therefore not possible in practice to discriminate to the satisfaction of all between the two kinds of isomerism which Berzelius had distinguished. Until general agreement could be reached on the formulae of the various isomeric substances, it was impossible to provide an explanation in other than extremely general terms. This did not happen until after the Carlsruhe Conference, when Cannizzaro's values were accepted. And at about the same time the theory was proposed that each atom had a definite capacity to combine with other atoms; its valency. Nevertheless, in 1864, William Odling, a distinguished theoretical chemist, said in his Presidential Address to the Chemical Section of the British Association that isomerism was still the problem of the day.

The study of isomorphism should have helped in this

situation, for if Mitscherlich's law were true, substances which crystallised in the same form would contain the same number of atoms. So the law could be used to compare atomic weights rather than equivalents. Unfortunately exceptions, some real and some apparent, to the law were discovered, and it fell into some disrepute once again; until it was rescued in the 1860s.

The problems which faced chemists in the understanding and application of the law of isomorphism are made clear in a long report presented to the British Association in 1837, with the title: 'on Dimorphous Bodies'. J. F. W. Johnston, the author, was a Fellow of the Royal Societies of London and Edinburgh, and had in 1833 been appointed Reader in Chemistry at the newly established University of Durham. Johnston remarked that the discovery of isomorphism had established that the crystalline form of a compound was not uniquely determined by its chemical composition. Even after this discovery, it was still believed that for any given chemical compound, the form was one and invariable. But this belief had been shattered by the discovery of dimorphism; sulphur, for instance, was found to exhibit two distinct crystalline forms, and the chemical identity of diamond and graphite had been established. These bodies were called dimorphous; and by then it was clear that crystalline form was neither specific nor constant. The dimorphous forms were found to exhibit no differences in chemical properties.

The complications continued to increase, for it was found that substances could be isodimorphous; that is to say, each could exist in two forms, in both of which the other could also be found to crystallise. Examples of this which Johnston provided are arsenious and antimonious oxides; and potassium nitrate and calcium carbonate. But this last example seemed to undermine the notion that even isodimorphism implied similarity of composition; for the formula of potassium nitrate was written (KO+NO$_5$), and that of calcium carbonate (CaO+CO$_2$). These formulae, now written KNO$_3$ and CaCO$_3$, and therefore in accord with the law of isomorphism, illustrate the chaos over atomic weights at this time, as well as the dualistic convention of writing formulae in two halves. Chemists could perform quite accurate analyses, and derive from them good values for equivalent weights; but they had no criterion, other than the arbitrary simplicity rules of Dalton, to guide them adequately in converting these equivalent weights to atomic weights. Atomic formulae which should have appeared analogous therefore looked quite different from one another, because the ratios of the atoms came out wrong. Thus in Johnston's formulae, some of the symbols stand for what we would call atomic weights, and others for equivalents.

The isodimorphism of calcium carbonate and potassium nitrate was only an illusory objection to the law of isomorphism, because their atomic constitutions, once Avogadro's law had been accepted, could be seen to be similar. More serious was the isomorphism between salts of potassium and of the ammonium radical, for the latter was known to be compound. In Johnston's words, potash (KO) was replaced

'by ammonia with an atom of water (H$_3$N+HO), without change of form, though neither the number of equivalents nor the number of elements, nor the ratio between the positive and negative constituents was alike in the mutually replacing compounds'.[67]

Johnston considered that the isomorphism between ammonium and potassium salts might be evidence for the complexity of potassium. We simply accept today that the radical ion ammonium, NH$_4$+, behaves very like the potassium ion, K+, although one is composed of five atoms and the other of but one; but in the 1830s this would have seemed like condoning a major breach in the law of isomorphism. Johnston spoke of those, carried away by Mitscherlich's beautiful doctrine, who had sought to reconcile the formulae; but he was himself tougher minded.

He went on to make the suggestion that isomerism and dimorphism might be connected. Carbon atoms differently arranged form diamond and graphite, whose physical properties are very different. Deeper-level rearrangements of the true atoms of bodies might produce substances differing in their chemical properties. In fact, the various elements might be related to one another as isomers are. This suggestion was taken up, and some transformations of one element into another were reported. In his popular *Familiar Letters on Chemistry*, Liebig asked rhetorically whether some of the elements might be mere modifications of one substance, the 'same matter in a different state of arrangement'; but he then proceeded to argue interestingly against the idea, on the grounds that the chemical elements are, in their various families, much more like one another in properties than true isomers should be. It had been found that the elements could be arranged in these families, all the members of which resembled each other quite closely; an example is the 'halogens'; chlorine, bromine, and iodine. If these were isomers, or polymers, they should be less alike; and in particular their compounds should not be isomorphous.

But this has taken us some way from our main object in this chapter, which was to show that there were good reasons why Avogadro's hypothesis was unacceptable to chemists; and that until these reasons had been shown to be less good than had been supposed, all the attempted extensions of the atomic theory into new fields did not prove fruitful. The phenomena of isomerism could not be explained in terms of atoms until an agreed scale of atomic weights, based not on hypothetical simplicity rules but on the combining volumes of gases, had been attained. And the same was true of isomorphism; the true and the apparent exceptions to Mitscherlich's law could only be separated and understood in the 1860s.

There were, however, other attempts to arrive at atomic weights from equivalents before that period. Gerhardt, a French chemist who believed that formulae should be mere recipes and not attempts to reveal hypothetical molecular structures, attempted to systematise the science. He based his atomic weights on the principle that the molecules of all elements in the vapour state are composed of two atoms. This solution, clearly the product of an excessively tidy mind, is in fact true of the then known gases of the atmosphere, but not of metals such as sodium, nor of phosphorus and sulphur. So Gerhardt's atomic weights, though definite, were not satisfactory either. And controversy continued over whether to have a 'one volume' or a 'two volume' system; that is, whether to suppose that the molecules of the elements are composed of one or of two atoms. Chemists of the 1850s continued to use 'equivalent' and 'atom' as synonyms; and the Carlsruhe Conference voted that the former was a more empirical idea than 'atom' or 'molecule'.

This purely systematic and theoretical solution by Gerhardt, an attempt to cut the Gordian knot, was not adequate to the needs of chemists. And a thoroughly empirical way of arriving at atomic weights from equivalents, that of Dulong and Petit, also proved delusive. Dulong and Petit had found that the product of the specific heat and the atomic weight of several elements is approximately constant, and has a numerical value of about 6·4. This constant product is called the atomic heat; and they declared their conclusion in the words: 'Atoms of all simple substances have the same capacity for heat.' Unfortunately evidence soon accumulated that the law was by no means universally true. The non-metals in particular did not obey it; and some clearly compound substances, for instance hydrogen peroxide, had the same value for their 'atomic heat' as typical elements. Hermann Kopp speculated in the *Philosophical Transactions* in 1865 on whether the smaller values for atomic heat which were found for the non-metals implied that their structures were simpler than those of the metals. The fact that the law clearly had exceptions, and that there was no unambiguous way of telling how many there were, meant that it was of little value in determining atomic weights until independent and unexceptionable methods of checking it had been found.

The state of feeling on the question of atomism is revealed in Mary Somerville's *Connexion of the Physical Sciences*, the ninth edition of which appeared in 1858. In the chapter on atoms and molecules, she mentioned Mosotti's theory, and also Rankine's theory of molecular vortices, according to which each atom consisted of an inappreciably small nucleus surrounded by an elastic etherial atmosphere. This theory could be extended to explain the liquefaction of gases by pressure. She then discussed isomorphism, and concluded that: 'substances having the same crystalline form must consist of ultimate atoms having the same figure and arranged in the very same order'. Then came Newton's argument for atomism, that unchanging atoms are required to explain the unchanging laws of nature.

When she came to discuss Dalton's theory, however, the learned author found herself in trouble, and had to admit that while Dalton's law—that bodies combine in definite and multiple proportions—'is fully established, yet instances have occurred from which it appears that the atomic theory deduced from it is not always maintained'. These instances seem to have been those gaseous combinations where half-atoms appeared to be involved. Water, according to Mary Somerville, was formed by an atom of oxygen uniting with an atom of hydrogen. In other words, she used equivalents instead of atomic weights.

In 1860 the Carlsruhe Conference was called to try to reach agreement on these problems. It seems that proceedings at the actual conference were somewhat disorganised, and little progress was made; but Cannizzaro had had the sagacity to have his paper, very ably recalling the chemical world to Avogadro's hypothesis, printed and circulated among the participants. On the train home several of them seem to have read it and become converted. Cannizzaro's paper consisted of lecture notes used by him in a course at the University of Genoa in 1858.[68] He began with an historical summary, giving some of the reasons why chemists had felt unable to accept the hypothesis, and showing how these had lost their cogency. Then he went on to show the hypothesis in action; and especially that from it a self-consistent series of atomic weights could be deduced, which fitted simply and easily into the structure of chemistry, without the anomalies which had been a feature of all the other ways of trying to find atomic weights from equivalents.

The hypothesis rapidly became accepted, and for the first time a series of agreed atomic weights was available. It was now possible to arrange the elements in a series of rising atomic weight in such a way that elements in the same family appeared near one another; the 'Periodic Table' of the elements, which adorns most chemistry lecture rooms today. This would have been impossible without atomic weights, for, because of the different valencies of the elements their equivalents do not form any such series. Indeed it now became possible to establish the notion of valency, the number of atoms of hydrogen (or similar element) with which one atom of the given element can combine, and to establish that this was definite, although some elements have two or more different valencies in different series of compounds. This was the first step towards an understanding of the structure of molecules, and towards the explication of such problems as isomerism. There was no longer any need for arbitrary rules, nor for wholly empirical laws known to have an open class of exceptions; and it began to be much more reasonable to think of formulae as representing atomic structures.

In short, after 1860, the use of a specifically atomic theory in chemistry had become much more defensible, for the theory now began to have distinct advantages in explanatory power over the idea of equivalents, and could be applied in a much more definite and certain manner than had hitherto been possible. It is therefore surprising to discover that in 1867 and 1869 the Chemical Society in London had two great debates upon the subject, in which the atomists came off worse. These debates, the last great triumph of the opponents of atomism in England, will be the subject of the next chapter.

6

The Debates

THE two possible attitudes towards chemical formulae were brought out clearly in a paper in 1850 by Alexander Williamson, who became Professor of Chemistry at University College, London; and who, as President of the Chemical Society, sought in 1869 to convince his fellow chemists of the usefulness of the atomic theory. In 1850 he pointed to the irregularity and inconsistency of the chemical formulae then in use; and reported Gerhardt's suggestion that the chaos might be resolved if formulae were regarded as simply 'synoptic', that is, providing a convenient summary of the reactions by which the substance is prepared. Williamson, who was no positivist although he had studied under Comte, was not prepared to allow that formulae were mere recipes; and he remarked that different formulae might result for the same compound if Gerhardt's programme were carried out, for many substances can be made in a variety of ways. Instead, or perhaps as well, formulae should be used: 'to represent what we conceive a compound *to be*, and should be such that it might be formed by any one of the various processes by which the compound may be prepared'.[69] The formulae should represent

> 'an actual image of what we rationally suppose to be the arrangement of constituent atoms in a compound, as an orrery is an image of what we conclude to be the arrangement of our planteary system'.

In 1850, when atomic weights and hence formulae could not be agreed on, this seemed a utopian programme. But with the coming of valency theory in the hands of Kekulé, Couper, and Frankland, and with the acceptance of Cannizzaro's atomic weights after the Carlsruhe Conference, it became much more reasonable to endeavour to make formulae represent structures. In 1867 an advertisement appeared in the journal *The Laboratory* for a set of balls and wires to construct molecular models, described as 'glyptic formulae'.

Positivists in the tradition of Wollaston were unable to accept happily such a state of affairs. The existence of atoms remained as hypothetical as ever, even though the atomic theory now began to exhibit what it had so noticeably lacked before, a cash value in chemistry. Mathematicians, notably Sir John Herschel, had continued to be uneasy at the progress of chemical notation; for chemical equations did not follow the ordinary rules of algebra. The formula for hydrochloric acid, HCl, implied that multiplication was somehow involved, for xy means x times y; so the formula should be written $H+Cl$. Again, no very obvious progress, so it seemed to Herschel, had been made towards making chemistry mathematical since the days of Davy and Dalton. And lastly, the influence of those who believed in the unity of matter still continued as strong as ever, both among disciples of Dumas and of Faraday, and among the physicists.

The three traditions of positivism, mathematical reasoning, and belief in the unity of matter were united in [Sir] Benjamin Brodie, the son of a President of the Royal Society with the same name. Brodie went to Germany to study under Liebig, and his earliest important theoretical publication was that mentioned earlier, in which he argued against the dualists that bonds between like atoms occurred. In this paper, which appeared in the same year as Williamson's, in 1850, Brodie already took a different line with respect to chemical formulae.

He believed, following Gerhardt, that a formula was but a memorandum of the reactions of a substance, and a mode of expressing its synthesis and analysis. He would not allow that it was 'a mode of indicating its molecular structure'. In a paper of 1864 Brodie reported on similarities between organic peroxides and chlorine; and the *Chemical News* mentioned that: 'Numerous tables of formulae were referred to in the course of the lecture'. These formulae were presumably to be interpreted as recipes.

By this time Brodie had held the office of President of the Chemical Society, and had become Professor of Chemistry in the University of Oxford, where he introduced laboratory teaching for students. In 1864 Odling, who was to be his successor at Oxford, reported in a speech to the British Association that Brodie was working on 'a new and strictly philosophical system of chemical notation by means of actual formulae, instead of mere symbols', and urged him not to delay in publishing it. In May 1866 Brodie read to the Royal Society an abstract of his Calculus of Chemical Operations, the first attempt to apply an operational calculus in the field of chemistry.[70] From Gerhardt he derived the idea that formulae should be considered as condensed recipes; and from George Boole a self-consistent algebra in which $x+y=xy$. Brodie's abstract attracted attention and produced a lively discussion, of which no record appears to have survived; and in 1867 he was invited to give a lecture on the subject of his Calculus before the Chemical Society. His title was 'Ideal Chemistry', and in his audience were a number of specially invited eminent physicists and chemists.

This lecture provides the clearest summary of Brodie's views that was ever published. The question of whether or not matter was atomic was, according to Brodie, not raised in the Calculus; the method was to be analogous to the application of algebra to geometry or to probability theory. The symbols used would not in any way resemble the objects symbolised, as Dalton had imagined his symbols did. Each symbol would be accurately defined, and every arrangement limited by fixed construction rules, whose propriety could be demonstrated. The Calculus, he had declared in 1866, was an 'investigation by means of symbols of the laws of the distribution of weight in chemical change'; and the end in view was: 'to free the science of chemistry from the trammels imposed upon it by accumulated hypotheses, and to endow it with the most necessary of all the instruments of progression, an exact and rational language'. In this last sentence he echoed Lavoisier and the reformers of chemical nomenclature of the eighteenth century. Brodie claimed as his most important predecessors in this objective the French chemists Laurent and Gerhardt, who had refused to accept the atomic theory, and had 'implanted in the science the germ of a more abstract philosophy, which it has ever since retained'.

Theory, Brodie told his audience of 1867, seemed essential to chemistry, for the science had only really begun with the phlogiston theory. Nevertheless, Davy had made his great discoveries without one, merely resting content with the facts of combination in definite proportions. Dalton's hypothesis was more audacious than the phlogiston theory, for it postulated that the observed continuity of matter was an illusion. In the sixty years of the existence of the hypothesis, chemistry had seen change but not progress; and that atomic models of balls and wires should now be on sale was the last straw. Chemistry must have gone off the rails of true philosophy when the atomic theory resulted in such a 'bathos'; a 'thoroughly materialistic bit of joiner's work', the culmination evidently of a whole series of errors and misconceptions. All this must be dropped, and chemists must instead adopt symbols with clear and distinct meanings.

In Brodie's system, all substances are considered in the

condition of perfect gases, under standard conditions of temperature and pressure; the 'unit' of ponderable matter is then that portion of it which occupies 1000 cc. On this 'unit', 'operations' are performed; and Brodie produced a glass cube and performed the operation of tapping it to illustrate this point. The symbol α stands for the operation by which a unit of space is turned into a unit of hydrogen; and other Greek letters stand for similar operations resulting in different substances. The symbols indicate not only the weight but also the kind of matter in a substance, rather as vectors indicate both size and direction. The various operations can best be imagined, according to Brodie, as packing; and the combinations of matter with matter and of matter with space are exactly analogous. Should anybody ask what chemical combination *is*, all that can be done is to give him a recipe. No metaphysical or atomic hypothesis could cast any light at all upon the question.

The weight of the unit of hydrogen was assumed to be unity; and hydrogen was also assumed to be a simple weight, represented by α and not α^2, which would be the formula were it necessary to perform α twice in order to produce hydrogen. This is equivalent to the assumption in the atomic system that hydrogen molecules consist of one atom, and Brodie's was therefore a one-volume system. The letter χ symbolised the operation which, performed upon a unit of hydrogen, yielded a unit of hydrochloric acid, the symbol of which was $\alpha\chi$ or $\chi\alpha$. Two units of hydrochloric acid yield one each of hydrogen and chlorine, containing the same ponderable matter. The equation for this change is:

$$2\alpha\chi = \alpha\chi^2 + \alpha.$$

The formula for chlorine is therefore $\alpha\chi^2$. Simplicity rules, akin to those of Cannizzaro, were provided to enable formulae to be calculated from data on the weights and volumes of the substances taking part in the reactions.

Using these rules, which mean that the simplest formula which is in accord with the facts should be accepted, Brodie deduced the formula $a\xi$ for water, and ξ^2 for oxygen. The symbol for hydrogen peroxide (H_2O_2 in our symbolism) was $a\xi^2$; this shows formal similarity to that of chlorine, $a\chi^2$, a similarity which would seem to reflect the similarity of properties between organic peroxides and chlorine which Brodie had earlier noticed. The unit of space, with symbol 1, enters the equation for the decomposition of hydrogen peroxide. The volume of the products exceeds that of the reactant by 50 per cent; so one unit of space and two units of hydrogen peroxide must have combined to yield three units of products, two of water and one of oxygen:

$$1 + 2a\xi^2 = \xi^2 + 2a\xi.$$

Brodie was able to produce equations, and formulae in this manner for a number of fairly simple reactions and substances.

An interesting point about the formulae generated for the elements was that they fell into three classes. Some like hydrogen were represented by a single Greek letter; in others the simplest formula required that two like operations be performed, as for oxygen, ξ^2; and others required two or more dissimilar operations; as chlorine, $a\chi^2$. No substance χ was known; and one might enquire whether the formula for chlorine implied that it was a compound of this unknown element with hydrogen. Brodie thought that probably it did, remarking that while his symbols did not necessarily imply the real existence of the things symbolised, they did not just come out of his head, but were 'ideal' in the sense that geometrical entities are. He suggested that substances such as χ might have existed in the primeval earth, but might no longer exist separately; a speculation which drew some support from contemporary ideas on the evolution of stars.

We should notice in passing that Brodie's predictions depended upon his selection of a one-volume system, to some extent at least. Had he chosen α^2 as his symbol for hydrogen, chlorine would have been χ^2, and there would have been no reason to suppose it a compound of hydrogen. But this aspect of the Calculus made it interesting to those who believed in the unity of matter, for Brodie seemed to be providing evidence for the decomposability of several elements, even though his system seemed to postulate a not inconsiderable number of genuine elements, or distinct operations.

The controversy which Brodie's lecture aroused was distinctly lively, and almost all the possible attitudes towards the atomic theory were expressed by some distinguished participant or other. In the debate which followed the lecture itself, the two most fundamental points were made by two physicists, Maxwell and Stokes. Maxwell suggested that the truth or falsehood of the atomic theory would be settled from dynamics, and referred to his own derivation of Avogadro's hypothesis from the assumptions of the kinetic theory of gases. When two distinct theories based on quite different evidence, physical and chemical, lead to the same conclusion, it would seem to be good evidence for their soundness. But chemists took no notice; Odling later told Sir Harold Hartley that chemists ignored physical evidence until after Cannizzaro's paper, and in fact this attitude seems to have persisted into the 1870s, when the union of the two sciences began to be cemented.

Stokes made the objection to which positivist science is always open, that the Calculus took into account existing knowledge and was indeed based securely on facts; but that new discoveries would make it liable to radical change. Such positivist constructions may turn out to be strait-jackets; and nature is usually less tidy than men like Brodie suppose. In fact Brodie had deserted his positivist faith in that he had made predictions, albeit hedged, that some elements would prove decomposable. But Stokes was right, in so far as it was found that the atomic theory could easily be made to explain the facts of isomerism, which could only have been included within the scope of the Calculus with great difficulty. The explanation of geometrical and optical isomerism given by van't Hoff and le Bel, in terms of the valence bonds of the carbon atom being directed to the corners of a tetrahedron, was ultimately very convincing. Brodie in 1879 asked for time to see how his Calculus could be extended to cover these phenomena, but time was not granted for he died the following year with the Calculus still unfinished. Nobody tried to complete it after his death; and later generations of chemists completely ignored the Calculus. In the history of the progress of the science it is of interest only because it forced the atomists to examine and clarify the assumptions which they had been making.

William Crookes published, in his journal *Chemical News*, a full text of Brodie's lecture, taken down in shorthand by a reporter; and also a summary of the discussion. It appeared under the title 'The Chemistry of the Future', with no question mark. In 1868 an article in the *North British Review* claimed that Brodie had made the question of whether the world was made up of atoms or not as open as it had ever been. Maxwell suggested to William Thomson that he should say something about the Calculus in his Presidental Address to the British Association, but this hint was not taken up.

To return to the discussion. The point which aroused the greatest interest amongst chemists was the hypothetical section in which Brodie predicted the decomposability of some elements. Roscoe evidently believed that these predictions were the most important part of the Calculus; for in dismissing it, in an unfair obituary of Brodie, he wrote that 'no experimental evidence was offered by him, and none of a satisfactory character was forthcoming'. It is strange that the whole point of a positivistic calculus should have been thus mistaken, Those who dissented from the Calculus attempted to show that it could be readily derived from the atomic theory, and was no more than an eccentric form of chemical notation. Williamson, who was in the chair at the meeting, declared that while he could not fully appreciate all the arguments used, he was sure that the publication of the Calculus would 'inaugurate an exceedingly important era in chemical language and notation'. Nevertheless, very shortly afterwards he read a paper to the Royal Society claiming that the Calculus was really based on the atomic theory.

Frankland, a pioneer of valency theory, declared that he did not really believe in atoms, nor in centres of force. Waterston, he said, had first protested against the present chemical symbols because they gave no idea of 'force'; an idea that would still be lacking if an operational symbolism were accepted. By 'force', chemical energy is meant; ordinary chemical symbols do not express the ease or speed with which a reaction takes place, nor the quantity of heat evolved or absorbed. Odling declared his amazement at what Frankland had said about the atomic theory; but he announced his disbelief too. The present notation, in his opinion, was based on an atomic theory whether one liked it or not; but the theory could be kept in the background. In books published in 1861 and 1870 his views can be more clearly discerned.[71] In 1861 he wrote of 'chemical atoms', and of atomic 'symbols' with plus and minus signs used 'almost in their ordinary algebraic sense' in chemical equations. Like Davy, he spoke of the 'proportional numbers' of the elements rather than of atomic weights. In 1870 he wrote that he deliberately avoided using diagrams of atomic arrangements after the manner of Kekulé. Such diagrams were occasionally useful, but their influence on the science was, in Odling's view, prejudicial, since they made 'the fanciful sticking together of variously pronged disks of more importance than the investigation of phenomena'.

In 1867 Kekulé published a paper 'On Some Points of Chemical Philosophy',[72] intended as an answer to remarks of this kind; Brodie referred on one occasion to Kekulé's 'scribbled pictures'. Kekulé wrote that while many chemists did not adhere to the atomic theory, all chemists used it:

> 'Some invent hypotheses with the view of obtaining the most satisfactory and consistent explanations of experimentally established facts, others develop the principles thus established to a greater extent, and thus construct a chemistry of formulae which runs side by side with the chemistry of facts, as something to a certain extent independent of it. Others again, though adhering to the same principle, keep to the form rather than the idea, and seem to think that the chemical constitution of a body is explained as soon as it is represented by a formula constructed according to the principles now in vogue.'

The atomic theory Kekulé believed to be metaphysical, and he did not himself believe in atoms in the literal sense. He thought nevertheless that the atomic theory should be used if it was useful; and chemical atoms indubitably did exist:

> 'The question whether atoms exist or not has but little significance from a chemical point of view; its discussion belongs rather to metaphysics. In chemistry we have only to decide whether the assumption of atoms is an hypothesis adapted to the explanation of chemical phenomena.'

The parallels between this passage and that of Lavoisier on the nature of the elements are striking. Lavoisier declared that speculations about the nature or number of the elements were metaphysical, and that chemists should therefore simply accept the elements as the limits of analysis. And Liebig had explicitly compared chemists' use of the terms 'atom' and 'element'. According to this view, chemical elements and chemical atoms might or might not have some connexion with the prime matter or matters from which the physicists believed the world to be made up. It was a point of view which was not destined to survive much longer, for physicists' atomic theories were beginning to invade the field of chemical theory; and whether chemists liked it or not, they would have to come to terms with physics.

Brodie's Calculus caused a stir at the British Association's meeting in 1867; where most attention was directed to the hypothesis that hydrogen was α, and to the predictions of the decomposition of certain elements—points which Kekulé had also noticed particularly in his essay. Then on June 3rd, 1869, Williamson, in his Presidential Address to the Chemical Society, sought to convince chemists of the truth and utility of the atomic theory.[73] His position was not very different from that of Kekulé, for he said that he was not interested in theories of matter but in chemical atoms; in fact he was defending an extremely weak form of atomism, as we shall see. His lecture differed considerably from Brodie's in its organisation, and was indeed reported to have been delivered from a sheaf of unsorted notes. Unlike Brodie, he failed to take every opportunity of revealing the disagreements and dissensions among his opponents. His objective was to show that those who wrote against the atomic theory nevertheless used it willy-nilly in their experiments and calculations. And also to commend a minimum, broad church, kind of atomism without unnecessary hypothesis; on which all must agree.

Williamson began by referring to the wide divergencies between what various distinguished chemists—he gave no names—said of the theory; and he noted that even in 1869 there was not yet complete unanimity on atomic weights, some authors still employing equivalents. He quoted from the *North British Review* the remark that the question of atoms was as undecided as it had been in Ancient Greece. Many chemists contrasted the 'hypothetical' atomic theory with the empirical laws of chemical combination; but in fact, said Williamson, many statements of these laws are as dogmatic as the atomic theory. This observation had been made over thirty years before by the Irish chemist Donovan, who had claimed that the laws of chemical combination contained just the same assumptions as the atomic theory. This could only be true when the atomic theory is made devoid of hypothesis, and becomes a theory of chemical atoms, which do differ very little, if at all, from equivalents. But we should note that they have a distinctly greater explanatory power in reserve, for it is possible to talk of different arrangements of chemical atoms, but not of equivalents or proportional numbers; and atomic weights were not equivalents, but incorporated some hypothesis, either that of Avogadro, or else of a one- or two-volume system.

Williamson's conclusion from this observed diversity of opinion, and from the similarity of the atomic and equivalents theories, was that

> 'on the one hand all chemists use the atomic theory, and that, on the other hand, a considerable number view it with mistrust, some with positive dislike'.

The theory should therefore be examined; and if it were found to be

> 'a general expression of the best ascertained relations of matter in its chemical changes, the only general expression which those relations have as yet found, and to be hypothetical only in so far as it presupposes among unknown

substances relations analogous to those discovered among those which are known, then it must be classed among the best and most precious trophies which the human mind has earned, and its development must be fostered as one of the highest aims and objects of our science'.

In this passage Williamson is clearly in the tradition which described by the term 'atomic theory' the laws of chemical composition; for he declared clearly that the only hypothesis in the theory was the inductive step from 'some substances combine in definite proportions' to 'all substances do so'. If this were all that he meant by 'atomic theory', then his audience would all have agreed with him, for all used the laws of chemical composition; but some might have preferred to employ a nomenclature which did not imply indivisible particles.

Williamson therefore sought to argue that the existence of atoms, in his minimal sense, remote from the Daltonian billiard-balls, is implied in the law of multiple proportions. All chemists, he said, used this law, and attributed any divergencies from it as due to experimental error. And analyses alone could not distinguish between for example $C_{27}H_{56}$, $C_{26}H_{54}$, and $C_{27}H_{54}$. This discrimination can only be achieved when we have a theory linking atomic structure and chemical character. Again, some elements have two or more different equivalent weights; but this can readily be explained on the atomic theory as a result of their variable valency. The atomic theory had (according to Williamson, though we may be permitted to doubt it) led to the development of molecular theories, and explained the gradation of properties in series of related compounds. The atomic theory could also explain the dynamics of chemistry. On the one hand, therefore, there is a theory which

'explains in a most consistent manner the most general

results of accurate observation in chemistry, and is daily being extended and consolidated by the discovery of new facts which range themselves naturally under it. On the other hand, we have a mere negation . . .'

Williamson did not mention Brodie's Calculus, which it would be unfair to describe in this way. Presumably he stuck to his notion that the Calculus could be derived from the atomic theory, and was equivalent to it, differing only in notation from the theory of equivalents.

Williamson ended his lecture with an appeal to put on one side all speculations as to the ultimate nature of matter; a very odd suggestion, to our way of thinking, from an eminent scientist lecturing on the atomic theory. Williamson based this appeal on the division which still existed between physics and chemistry; theories of matter came under someone else's province. He declared:

'The question whether our elementary atoms are in their nature indivisible, or whether they are built up of smaller particles, is one upon which I, as a chemist, have no hold whatever, and I may say that in chemistry the question is not raised by any evidence whatever. They may be vortices, such as Thomson has spoken of; they may be little hard indivisible particles of regular or irregular form. I know nothing about it; and I am sure that we can best extend and consolidate our knowledge of atoms by examining their reactions, and studying the physical properties of their various products, looking back frequently at the facts acquired, arranging them according to their analogies, and striving to express in language as concise as possible the general relations which are observed among them.'

This denial by Williamson of interest in theories of matter

might have been expected to reconcile the atomists and those who adhered to belief in the unity of matter.

Williamson might well have regarded this degree of separation between chemistry and physics as necessary to the progress of the former science. For if chemists had had to wait until physicists had produced an atomic theory competent and detailed enough to explain the facts of chemistry, then the science could not have begun to make progress until at least a century after the time of Lavoisier. Chemists, using chemical evidence alone, had by 1869 arrived at an atomic and molecular theory having wide explanatory power within their own sphere. But by this time it would seem that this autonomy should have been becoming a thing of the past; it was already an old-fashioned view. Maxwell had shown how the kinetic theory of gases agreed with the molecular theories of chemistry; and there had been throughout the century numbers of chemists who were not prepared to accept a chemical atomic theory, postulating 'chemical atoms', unless it were grounded in a theory of matter. Brodie had pointed in the opposite direction; because the theory of chemical atoms was not a theory of matter, but looked rather like one, and had therefore loose connexions with physics, he believed that it would be more honest to drop it, and exclude all such vain questions from chemistry for ever by enclosing the science in the strait-jacket of the Calculus; in which descriptions of phenomena, put into mathematical terms, and wearily repeated to persistent questioners, passed for explanations.

But we must return to the discussion of Williamson's lecture, which came later in the year 1869, on November 4th.[74] The accounts which we have of this debate are highly compressed, and according to Williamson's obituarist they 'give no adequate reproduction of what was said . . . excitement ran high at times'. Brodie took the chair, and introduced the discussion by remarking that since there were several kinds of atomic theory in chemistry, it would be well to decide at the outset which one was to be discussed. Williamson denied this, arguing that Dalton's was the germ of an atomic theory, which had since developed. Brodie had also alleged that Williamson in his address had failed to distinguish fact from fiction; a charge Williamson also denied.

The debate was now open, and Frankland began it by saying that he was not a blind believer in atoms. He agreed with Williamson's contention that everybody used the theory, but suggested that therefore the only point of real interest was its truth. Since this was 'unknowable', the best thing to do was to use the theory as a 'kind of ladder to assist the chemist', but to be prepared, like Faraday, to put it on one side on occasion. The theory was perhaps for him a useful fiction; he would have been prepared to allow theoretical entities in the sciences provided they were not taken too seriously. Miller, a pioneer of stellar spectroscopy, also believed the theory to be a useful fiction. In order to reason about the facts, he claimed, one had to accept some kind of hypothesis; physicists used the undulatory theory of light, and chemists should use the atomic theory in the same manner. This was the parallel which had been drawn by the author of the article in the *North British Review*.

Both these men, then, agreed that the theory was necessarily but a useful fiction, but disagreed as to the extent of its utility. Odling took a somewhat tougher line than Frankland; he was not, he said, very interested in whether atoms existed or not, and Williamson had certainly not proved that they did. He believed that chemists did not base their researches on the atomic theory, but on the empirical law of multiple proportions. He recalled the audience to Davy's remarks on the theory; the laws of chemical combination were to be compared to Kepler's laws, general expressions of observed facts. Hypothesis was not necessary. We should record that Odling later became an atomist, and declared in 1898 that he had always followed in Williamson's footsteps, but had sometimes lagged behind.

John Tyndall, a physicist and populariser who found few difficulties in belief in theoretical entities, supported Williamson. He remarked that Newton had followed Kepler; and suggested that the atomic theory was comparable to the theory of gravity. Even the best established theories, such as the emission theory of light had seemed to be, might fail, and so the atomic theory in chemistry might prove inadequate; but as long as a theory explains the facts it will stand. Tyndall reaffirmed this the following year, in an address which he began with a quotation from Brodie's father on the use of the imagination in science. He said:

> 'Many chemists of the present day refuse to speak of atoms and molecules as real things. Their caution leads them to stop short of the clear, sharp, mechanically intelligible atomic theory enunciated by Dalton, or any form of that theory, and to make the doctrine of "multiple proportions" their intellectual bourne. I respect the caution, though I think it is here misplaced. The chemists who recoil from these notions of atoms and molecules accept, without hesitation, the Undulatory Theory of Light. Like you and me they one and all believe in an ether and its light-producing waves.'[75]

There were apparently two more major contributions to the debate. Carey Foster, a physicist, argued that atomic explanations were not necessary in chemistry; that chemical changes could be looked at in quite another way. And Edmund Mills declared his belief in the ultimate continuity of matter. Carey Foster had been a pupil of Williamson's, but had transferred his interests, and was now Professor of Physics at University

College, London. He admitted that the atomic theory was useful, but pointed out that the history of science was littered with false theories which had in their day been useful. He then proposed a kind of chemical explanation, akin to that of Coleridge, quite different from those used by atomists or by those who employed equivalents. It is not clear whether he believed his own explanation, or was proposing a different scheme to make the logical point that facts never demand one and only one theory to explain them; indeed in debates of this kind to disentangle the logical and scientific points is often hard. As Foster said, in thinking of a compound we naturally imagine the components occupying different portions of space. But combination might be thought of as a kind of transmutation:

> 'We know that between the bodies which disappear and the body which appears, there are certain relations, not only qualitative but quantitative, the total mass of the disappearing substance being equal to that of the appearing substances; but we may perhaps return, sometimes at any rate, with great benefit, to the notion that one portion of matter is actually transmuted into another; that it ceases to exist as such, but something else comes in place of it. From such ideas the existence of atoms would not follow of necessity, but with our present mode of stating and reasoning about chemical changes, an atomic hypothesis or basis appears to be inevitable.'

Brodie's Calculus might perhaps be viewed as an example of the kind of chemistry Foster had in mind; for the mysterious 'operations' do produce a kind of transmutation which cannot be further investigated. And one of the reasons why Brodie had evolved the Calculus, rather than continue to use equivalents as Wollaston had done, was that he believed that from equivalents to an atomic theory the passage was too easy. But it seems more probable that Foster was in touch with the German tradition of *Naturphilosophie*, as Coleridge and Faraday had been, and was trying to recommend a more metaphysical chemistry than Brodie would have liked. We can sympathise with those who believed that the whole is greater than the sum of its parts, and who sought a chemistry based on transmutation of bodies having a certain set of qualities into others with different qualities. But it is hard to see how a science of chemistry could have been constructed along these lines, at any rate as speedily as our science of chemistry was erected; and clearly it would have been a very different chemistry from ours.

Edmund Mills' contribution to the debate was also radical and metaphysical in character; and his contemporaries seem to have felt that he was too much the metaphysician to be a trustworthy theoretical chemist. Mills pointed to the absence of sharp boundaries in nature; and to all the force and motion in the world, of which the atomists took no notice. All the sciences, he believed, converged to some limit; like scientists of the generation of Davy and Dalton, he longed for a law from which all phenomena might be deduced, but in its absence thought that there was one general idea against which all theories should be measured. This idea was *motion*, or approximately, in our terms, energy.

In order to understand Mills' point, which is lost in the compressed report of the debate, one must refer to the series of articles which he published in the *Philosophical Magazine* in 1869, 1870, 1871, and 1873. The third of these refers particularly to the atomic theory, and is the most interesting therefore for our present purposes.[76] Most chemists, Mills remarked, had followed Dalton in thinking of atoms as essentially static; a point of view exemplified in the wires holding the wooden balls together in glyptic formulae, and the lines or bonds connecting the atoms in graphic formulae. No chemical discoveries had cast any light on the question of whether or not atoms were indivisible; nor had they proved the existence of atoms. No atomic theorists could give any satisfactory explanation of affinity; nor had they made any headway in elucidating isomerism:

> 'Isomerism is not . . . explained by assertions about indivisibles, which have neither been themselves discovered nor shown to have any analogy in the facts or course of nature— nor by explicit statements about a "structure" which has never been seen—nor by the use of a phrase to which no clear definition has been, or can be, attached.'

Definite proportions are all that are expressed by chemical equations, according to Mills, and definite proportions do not require an atomic explanation. The continuity of the states of matter, the hazy boundary between the liquid and vapour states revealed by Andrews' experiments, for example, implied for Mills the continuity of matter. We jump, he wrote, at absolutes because of our impatience at the recondite nature of forces which we cannot see:

> 'Surrounded on all sides with continuity, motion, and change, our most popular ideas relate to limits, repose, and stability.'

The atomic theory fits easily into the prejudices of our education; even for a young Idealist, it is difficult to be content with Nature as she is. Nobody has ever seen an atom; its status is the same as that of phlogiston. The nature of phlogiston varied, in the course of time, as much as the atomic weights of the elements had; but even phlogiston never had attached to it a property without natural analogy, as atoms have in their alleged indivisibility. The atomists must say with Tertullian: 'Certum est, quia impossibile.'

After some more mildly spiteful remarks about atomists, Mills passed on to review the opinions on the question of a number of earlier scientists of the first rank. Newton's mind was not quite made up, he concluded; and Descartes, Liebniz, and Kant he enrolled among the non-atomists. Davy and Wollaston had opposed the theory; and so had Faraday, who, in his support for the atom of Boscovich, had proclaimed himself at once an idealist and a Leibnizian. No satisfactory answers had ever been made to his objections to the atomic theory in its usual form. In later papers, Mills embraced the doctrine of the unity of matter as well as its continuity; and appealed for someone to finish Brodie's Calculus.

The debate concluded with Williamson declaring that he would rejoice if a better theory were produced to replace the present one; and Brodie castigating those who had held that the atomic theory was a useful fiction. He could not, he said, understand using a theory and denying it at the same time. Preferring clear and definite ideas, he declared himself incapable of forming the idea of an atomic theory in the abstract; but he believed that the atomic theory in chemistry was harmful and should be dropped. The books of Kekulé and his disciples, 'scribbled over with pictures', were mischievous and misleading to students, who might think that such things were fundamental.

To contemporaries, the debate brought out with great clarity the divisions between chemists which had for fifty years been more or less concealed behind such phrases as 'the atomic theory in a form devoid of all hypothesis'. Realists, positivists of various hues, and believers in a continuum were all brought face to face and made to thrash out their differences. One cannot help feeling that it was the opponents of atomism who came off best in the discussion, even though Williamson was trying to present a case for a minimal atomism. Reporting the debate in an editorial, the new journal *Nature* remarked on the gulf, which would perhaps prove impossible to bridge, between the participants. In the next few years there were other controversies over the atomic theory before the Chemical Society; but little new was added, for after the debates of 1867 and 1869 there was little left to say. Meanwhile, organic chemists, armed with the atomic theory, continued to subdue their difficult terrain, and to produce a series of triumphs, without much reference to the question of the actual existence of atoms.

The success of structural theory, first in organic chemistry and then in inorganic—in elucidating the complex ions of the transition metals—ensured that atomism would have to be taken more seriously, and some of its opponents, such as Odling, were in time converted. It is therefore odd to find that in 1907 Williamson's obituarist[77] recorded of the 1869 affair that 'nothing came of it all, and chemists remain not much less divided on the subject now than they were then'.

But very shortly after these lines were written, the resistance of the sceptics collapsed with the conversion of Ostwald to atomism. In the final chapter we shall be concerned with this last phase of scepticism, and with the changes which led to the abandoning of the indivisible atom in favour of a particle having structure. And also with the increasing co-operation between chemists and physicists as the gulf between the two sciences closed with the attainment of an atomic theory broadly satisfactory in both disciplines.

7

The End of the Affair

THE successes of structural theory in organic chemistry could of course only add to the usefulness of the atomic theory in chemistry; they could not prove the existence of indivisible particles. But whereas in the 1860s it was still a possibility that a non-atomic chemistry would prevail, by the 1880s the difficulties of providing an explanation of optical isomerism that was as adequate, simple, and consistent as that supplied by the atomic theory were enormous. In 1887 Williamson, in a farewell speech at University College, London, made much the same points as in his address nearly twenty years before to the Chemical Society. He said:

> 'A scientific theory, I suppose, ought to be the most condensed statement of general facts: and I take it that the atomic theory is that. To me it is nothing more. I do not use atomic reasoning, or refer to considerations relating to atomic operations, otherwise than as indicating limits which exist to our efforts to break up matter. Whether we like them or not they exist. There are thinkers who prefer overstepping that boundary and discussing what there may be beyond it—whether, for instance, these limits by which we are stopped at present, and which I think it is well to recognise as existing in our present state of knowledge may not be overstepped. As a chemist I have nothing to do with that.'[78]

In this passage, Williamson has moved a little way from simply arguing that the atomic theory is useful; but if the points he made had been the only ones at issue, and if all chemists had been able to ignore physical theories, then there would have been little controversy in 1867 and 1869, and certainly none later in the century.

In a book written, it seems, in the second decade of the present century, for the Home University Library, one finds perhaps the last appearance of the minimal, chemical atomism of Williamson and of the writers of textbooks in the mid-nineteenth century. In its extreme form, wrote Professor Meldola,[79] the atomic theory had

> 'been pushed so far as to claim that the atom is the ultimate component of the material universe—imperishable and eternal. The reader must, however, discriminate between ascertained concordance of facts with theory and speculative developments of theory, however plausible in principle or stimulating in prompting further research.'

The discussion of the truth of atomism belonged to Philosophy; chemists had only to deal with the doctrine in so far as it was in accordance with chemical facts. In this sense the atomic theory was the headstone of theoretical chemistry; but the mysteries of chemical change were not explained by it.

Meldola considered, correctly as it turned out, that 'the ultimate coalescence of Physics and Chemistry' would be brought about through an interpretation of the Periodic Table. In fact this coalescence had been happening slowly since the 1870s. In 1870, the chemist Henry Roscoe,[80] referring at the British Association's meeting to the debate of 1869, agreed that the atomic theory was not necessary in chemistry, though it could explain definite proportions and isomerism. But he noticed William Thomson's conclusion, from physical evidence, that matter is discontinuous, and also the kinetic theory of gases, the success of which made 'the existence of indivisible particles more than likely'. Roscoe was one of the first chemists to take note of physical evidence in this way; but in the next few years more and more began to think like him.

In 1871 Tait, the physicist, told the British Association that there could be little doubt but that the second law of thermodynamics contained implicitly the whole theory of chemistry, allotropy, fluorescence, and thermoelectricity; a notion which reappeared in Continental authors who sought a basis for chemistry in thermodynamics rather than in atomism. In 1874 Jellet in the Physical Section and Crum Brown in the Chemical, suggested that chemistry would come under mechanics. Crum Brown, who had earlier proposed a calculus like Brodie's but incorporating atoms, said that 'we are struggling towards a theory of Chemistry. Such a theory we do not possess.' And we shall not 'until we are able to connect the science by some hypothesis with the general theory of Dynamics'. Jellet argued against Comte that chemists should not seek to defend the autonomy of their subject, and pointed out how valuable had been the introduction of mathematics in optics.

The addresses of some Presidents of the Chemical Section during the 1880s are of extraordinary interest for the light they cast on the attempt to join physics and chemistry by some kind of atomic theory, and also for the ideas they contain on the nature of the elements. During this epoch the various separate threads which we have been following began to be woven together, and the story starts to move at breakneck speed. In 1882, Liveing made the transition from an appeal for a dynamical chemistry to speculation on the nature of the elements. The most important recent advance, he said, had been 'the attempt to place the dynamics of chemistry on a satisfactory basis, to render an account of the various phenomena of chemical action on the same mechanical principles as are acknowledged to be true in other branches of physics. I cannot say that chemistry can yet be reckoned among what are called the exact sciences . . . but that some noteworthy advances have in recent years been made.'

If all energy were one and indestructible, then there could be only one set of dynamical laws; and these must apply in all branches of knowledge.

Textbooks should, he believed, show more clearly the importance of the concept of energy. Molecular diagrams were merely an aid for schoolteachers. The kinetic theory of gases and the vortex atom had helped to loosen the hold on the imagination of the Epicurean hard atom, with its atmosphere of heat, which had always presented especial difficulties. The notion that bonds between like atoms were impossible had been exploded; but, he asked,

'why stop here? Why may not the chemical elements be further broken up by still higher temperatures? *A priori* and from analogy such a supposition is extremely probable.'

The notion that there was but one basic matter had had a long history, but by this time it was usually associated with the name of Prout. His hypothesis, that hydrogen might be this matter, had not survived the proof that atomic weights are not all whole numbers on the scale $H = 1$; but later Proutians, such as Dumas and Marignac, had simply postulated a lighter first matter. Marignac had also remarked on how close many atomic weights were to whole numbers, far more than could be due to chance; and Prout's hypothesis, in one form or another, was still felt to be an open question. Liveing felt that the complexity of the spectra of some elements proved that they could not be simple. These were the spectra produced when a substance is vapourised in a flame of arc, which is examined through a spectroscope. The spectrum consists of bright lines, characteristic of each element; and iron, for example, has a great many lines in its spectrum; therefore, in Liveing's view, it was probably complex.

In 1884 Lord Rayleigh described how he had tried to draw the attention of chemists to the second law of thermodynamics; and declared that the success of the kinetic theory showed that some at least of its fundamental postulates were in harmony with the reality of nature. In 1882, he had mentioned Prout's hypothesis, noting that some chemists opposed it *a priori*, while others felt that its simplicity justified its acceptance. Rayleigh believed that it was time once again to measure gas densities accurately, and calculate atomic weights from them to test the hypothesis; a research programme which led ultimately to the discovery, in co-operation with Ramsay, of argon, the first of the inert gases of the atmosphere to be isolated.

In 1883 J. H. Gladstone, a successor of Faraday, as Fullerian Professor at the Royal Institution, gave an address devoted to the problem of the elements. Chemists of his generation, he said, were tempted to look upon the elements as genuine simple forms of matter; but the question demanded careful examination. Spectroscopy did not reveal lines due to some common real element in the spectra of various elements; but stellar spectroscopy did seem to have produced some evidence for the dissociation of elements at high temperatures. The radicals now seemed less like elements than they had, for there was no room for them in the Periodic Table. And the atomic weights of elements did not show, in their families, the regular increments to be found in the homologous series of organic chemistry, though Dumas had tried hard to find such relationships. Having thus demolished all these empirical arguments, or at least shown that they wanted cogency, Gladstone rather oddly produced a number of reasons for favouring a Proutian scheme:

> 'There does not seem to be any argument which is fatal to the idea that two or more of our supposed elements may differ from one another rather in form than in substance, or even that the whole seventy are only modifications of a prime element; but chemical analogies seem wanting.'[82]

The only satisfactory analogies would seem to have been close parallels between the elements and radicals; for Gladstone went on to add:

> 'the remarkable relations between the atomic weights of the elements, and many peculiarities of their grouping, force upon us the conviction that they are not separate bodies created without reference to one another, but that they have been originally fashioned, or have been built up from one another, according to some general plan.'

Crookes in 1886 set out to supply this plan, and took over from biology the theory of evolution. But this was by no means the first application of evolutionary views to the chemical elements, for Asa Gray, in his famous review of the *Origin of Species*, had remarked that sympathy with theories of inorganic evolution might ease the acceptance of Darwinism. Maxwell thought that the application of such theories to atoms and molecules was absurd; unlike organisms, they are not born, they do not die, and all of a kind seem identical. Maxwell and Herschel believed that the only sensible analogue to a molecule was a manufactured object; a shilling from the Mint, in Stokes' gloss.

Crookes' theory of evolution was not open to these objections, for it was a form of creative evolution, following a plan; and moreover his atoms were not all alike. He set off with a splendid purple patch:

> 'Inquirers, working at the very confines of our knowledge, find themselves occasionally at least face to face with a barrier which has hitherto proved impassable, but which must be overthrown, surmounted, or turned, if chemical science is ever to develop into a definite, an organised unity. This barrier is nothing less than the chemical elements commonly so called, the bodies as yet undecomposed into anything simpler than themselves. There they extend before us, as stretched the wide Atlantic before the gaze of Columbus, mocking, taunting, and murmuring strange riddles which no man yet has been able to solve'.[83]

All definitions of the term 'element', even operational ones, were unsatisfactory, said Crookes, because they depended upon the limitations of human power. Quoting from Faraday, Lockyer, Stokes, Brodie, Graham, and Herbert Spencer as his authorities, he was able to declare that the notion that the elements were really compounds was 'in the air of science'. Probably, he opined, 'there exist in Nature laboratories where atoms are formed, and laboratories where atoms cease to be'. The array of elements in the Periodic Table resembles a biological classification. Some groups have many 'species', and some few; some species are common, and some rare; some whole groups are of restricted occurrence. The rare earth elements in Sweden might be compared with the marsupials in Australia. Evolutionary explanations which fit the organic world might therefore be expected to apply to the inorganic also; and there might even be an inorganic analogue to the struggle for existence.

> 'If evolution be a cosmic law, manifest in heavenly bodies, in organic individuals, and in organic species, we shall in all probability recognise it, though under special aspects, in those elements of which stars and organisms are in the last resort composed.'

It seemed likely that as the world cooled heavier and heavier elements were formed; and Crookes spoke of a pendulum which swung back and forth, not quite covering the same path at each swing. Thus at a given point on the first swing lithium would have been formed, on the second sodium, and on the third potassium. A three-dimensional, figure-of-eight form of the Periodic Table was made by Crookes to illustrate this. Inorganic evolution might well by now have stopped, as many biologists believed that organic had. The atoms of elements were probably of different weights, grouped around an average value. His own researches on the rare earth metals, which are extraordinarily hard to separate because of their close chemical similarities, had convinced him that they were not really separate bodies but a continuous group. On the other hand, atoms did not differ in weight continuously, but *per saltum*; Crookes seems to have believed that the weights of all atoms were exact multiples of that of hydrogen. Repeated fractionation among the rare earths was not really separating numerous distinct elements, but heavier and lighter atoms of yttrium and one or two other elements. Slower cooling of the rare earths would have led probably to the formation of only one element; but as it is these minerals are

> 'the cosmical lumber-room where the elements in a state of arrested development—the unconnected missing links of inorganic Darwinism—are finally aggregated'.

To such living fossils the normal rules of classification could not with precision be applied.

The characteristic of each element was not its atomic weight but the quantity of electricity which was bestowed upon it at the moment of its birth; the quantity on which its valency depended. This notion that electricity might be atomic—communicable only in discreet units—was proposed by Johnstone Stoney in the 1870s and again, independently, by Helmholtz in 1881 in his Faraday Lecture.[84] Helmholtz drew attention to Faraday's law of electrolysis, that the same quantity of electricity liberates equivalent weights of the elements, and added:

> 'If we accept the hypothesis that the elementary substances are composed of atoms, we cannot avoid concluding that electricity also, positive as well as negative, is divided into definite elementary portions, which behave like particles of electricity.'

Helmholtz also gave general support to the idea that chemical affinity was electrical in nature, recommending a distinction between two kinds of compounds, typical ones and 'molecular aggregates' in which the bonding was of some other kind.

At about the same time Crookes had discovered what he described as the fourth state of matter; the radiation from the cathode, or negative terminal, in a tube which was almost evacuated. Crookes discovered that the rays went in straight lines; and that the material from which the cathode was constructed made no difference. He identified these 'cathode rays' therefore as the prime matter from which everything was made. But he had a different candidate for this prime matter in his lecture to the British Association in 1886; the element helium, identified spectroscopically in the sun by Lockyer, and present there in large quantities. No substance having this spectrum was known on earth; and it therefore seemed possible that all the terrestrial helium had, as our planet cooled, been transformed into heavier elements. About the turn of the century helium was discovered, occluded in some minerals, by Ramsay, and found to have an atomic weight greater than that of hydrogen; so it ceased to be a candidate for prime matter after all.

In 1889 Crookes elaborated before the Chemical Society his ideas on the composition of the rare earth elements. The spectroscope, he said, 'enables us to peer into the very heart of nature',[85] and it threw light 'upon the nature and the relations of our *elements*, real or supposed'. As the earths were fractionated, great numbers of claimants for the status of element arose; and the problems were whether they could be fitted into the Periodic Table, and whether they could be clearly defined. The spectra of these bodies were not wholly distinct, but shaded into one another; and Crookes' suggestion was that the rare earth elements must be considered as composed of certain simple substances, called 'metal-elements'. He declared:

> 'Our notions of a chemical element must be enlarged; hitherto the elemental molecule has been regarded as an aggregate of two or more atoms, and no account has been taken of the manner in which these atoms have been agglomerated. . . . We may reasonably suspect that between the molecules we are accustomed to deal with in chemical reactions, and the component or ultimate atoms, there intervene sub-molecules, sub-aggregates of atoms, or meta-elements, differing from each other according to the position they occupy in the very complex structure known as "old yttrium".'

These suggestions of Crookes' do not appear to have had any direct role in the formation, in the twentieth century, of the theory of isotopes; this theory was devised to explain quite different phenomena. Again, his theory of matter seems vague, and not much closer to verifiability than Davy's suggestions of sixty years before. On the other hand, Crookes' theory of the evolution of the elements has found its way, in modified form, into modern chemistry. And his work on cathode rays paved the way for J. J. Thomson, for the identification of the cathode rays and the atoms of electricity as 'electrons', and for the atomic physics of the turn of the century.

J. J. Thomson's classic paper of 1897 contained his estimate of the ratio of charge to mass of the corpuscles which composed the cathode rays[86]. German scientists had believed the rays to be etherial undulations like light; but Thomson had preferred a corpuscular theory because it was definite, and led to predictable consequences. He deflected the rays both by a magnetic field, and also by an electrostatic field—an effect which none of his predecessors had achieved. He concluded:

> 'As the cathode rays carry a negative charge of electricity, are deflected by an electrostatic force as if they were negatively electrified, and are acted on by a magnetic force in just the way in which this force would act on a negatively electrified body moving along the path of these rays, I can see no escape from the conclusion that they are charges of negative electricity carried by particles of matter.'

Two independent methods of estimating the ratio of charge to mass indicated that it was over 1000 times greater than that for the hydrogen ion. The particles must therefore either be extremely small, or carry an enormous charge.

Thomson plumped for the conclusion that they were very small, though he also thought that they probably carried a larger charge than the hydrogen ion. The rays were always the same whatever gas was in the tube, and there seemed good reason for believing that their mass was genuine and not a quasi-mass due to the field which a charged body sets up. Thomson inclined to the view that they were the prime matter of the chemists:

> 'The explanation which seems to me to account in the most simple and straightforward manner for the facts is founded on a view of the constitution of the chemical elements which has been favourably entertained by many chemists: this view is that the atoms of the different chemical elements are different aggregations of atoms of the same kind.'

Thomson had been attracted by this aspect of the vortex atom, and must have been pleased to see his experiments justify his preconceptions in so precise a way.

The cathode rays seemed to be the prime, basic state of matter, and the problem was to fix on stable configurations in which they could compose the 'atoms' of elements. Some law of force was required—perhaps that of Boscovich—but the problem seemed incapable of general mathematical solution because of the number of corpuscles involved. But experiments with floating magnets showed that they arranged themselves around a central magnet in a series of rings, which seemed to show some analogies with the periods of the Periodic Table. We should notice that in this new corpuscular philosophy, as in the old one of Newton and Boyle, transmutation remained a reasonable objective. And indeed at about the time when Thomson was performing these investigations, researches on radioactivity were proceeding which did in time lead to transmutations.

Crookes in 1902 remarked that what he had twenty-five years before called 'radiant matter' now passed as 'electrons'. He held a nuclear theory of the atom, believing that a chemical ion was 'a material nucleus or atom of matter constituting by far the larger portion of the mass, and a few electrons or atoms of electricity'.[87] The more usual view seems to have been like that put by Sir William Tilden in 1910; that the atoms of the elements are composed of corpuscles, all negatively charged, and therefore held together by the presence of an equivalent amount of positive electricity. No positive particle with a mass less than that of the hydrogen ion had been discovered; and Thomson had in 1904 proposed an atom in which negative particles were distributed within a sphere of positive electricity of uniform density—the plum pudding atom. This atom was probably the nearest return to the theories of the corpuscularians that was to be seen; for since that date atomic models have become increasingly complicated, and the number of sub-atomic particles is now as scandalous as the number of elements was in the nineteenth century.

The study of cathode rays made it no longer possible to adhere to a simple Daltonian atomism, and also made the vortex atom no longer of interest. But it did not force all chemists to believe in the real existence of atoms any more than had the successes of structural theory in organic chemistry. In 1898 F. Wald wrote an article in the *Journal of the Chemical Society*, seeking to show that there was no necessity for the assumption of an atomic hypothesis. He based definitions of similar and dissimilar materials, and of a chemical compound, on the Phase Rule, a theory-free rule connecting the number of components in a system; the number of phases [gas, liquid(s), and solid(s)]; and the 'degrees of freedom' of the system, the changes of temperature and pressure it can undergo without upset in the equilibrium. Wald claimed that chemists had always been guided, perhaps unconsciously, by some such principle in their analyses.

In 1901 Professor Rücker, in his Presidential Address to the British Association,[88] sought to fix the attention of his hearers on the hypotheses and assumptions upon which 'the fabric of modern science has been built, and to enquire whether the foundations have been so "well and truly" laid that they may be trusted to sustain the mighty superstructure which is being raised upon them'.

Among the three basic assumptions which he chose was the atomic theory. Positivists urged, he said, that hypotheses were simply useful fictions; and models of only secondary importance. The atomic theory seen in this light was unnecessary, an attempt to give precision to ideas which must be vague:

> 'A great school of chemists, building upon the thermodynamics of Willard Gibbs and the intuition of van't Hoff, have shown . . . that if a sufficient number of the data of experiment are assumed, it is possible . . . to trace the form of the relations between many physical and chemical phenomena without the help of the atomic theory.'

But such a surface treatment could not, according to Rücker, be the last word. Positivists like Stallo would no doubt have retorted that a surface treatment was greatly to be preferred to a pseudo-explanation, in terms of entities of which we know nothing.

Rücker did not accept that we can know nothing of atoms, for Saturn's ring, which was even more inaccessible, was generally believed to be composed of particles. Some coarse-grainedness of matter was indicated by the way gases diffused into one another, and by the interpenetration of solids; observed, for example, in gold and lead which have been pressed together for years. Further, no clear picture of the phenomena of heat could be given without some kind of particulate hypothesis. Rücker distinguished two sorts of atomic theory, in one of which the atoms were different in kind from the ether in which they swam, whilst in the other they were parts of that medium—vortices, for example. In both cases the atoms seemed to have properties different from those of ordinary matter, and no wholly satisfactory description of atomic constitution seemed yet to be available. On the other hand the atomic theory imparted 'a unity to all the physical sciences which has been attained in no other way'. The cloud chamber, in which charged particles cause water vapour to coalesce into droplets, seemed strong direct evidence for molecules; and so did Lord Kelvin's estimates of the sizes of atoms, calculated from a range of different phenomena. Rücker, in the manner of Williamson, challenged opponents to produce a better theory if they could.

Most of his evidence was evidence only for the existence of molecules, and it would be quite possible to have these without atoms. The chief difference between the speeches of Rücker and of Williamson in 1869 was that Rücker argued not only for chemical atoms but also for physical atoms or molecules, whose sizes and properties were consonant with the needs of chemistry. It is surprising to find van't Hoff, one of the founders of structural organic chemistry, appearing among the physical chemists for whom atomism was an unnecessary hypothesis. At this time it still appeared possible that chemistry and physics might be united through thermodynamics; a mathematical treatment based on empirical data, in which the question of atoms was not raised.

The doubts of Rücker contrast strongly with the confidence of Tilden, writing nine years later in 1910. Modern experimental physics and chemistry, he declared, had transformed the atomic theories of Antiquity:

'The old question whether the divisibility of matter is finite or infinite, debated throughout ancient and medieval times without the possibility of reaching a conclusion, has been in modern times decisively answered by physics and chemistry. It is no longer a subject of debate. . . .'[89]

The main evidence was, on the physical side the phenomenon of diffusion; and on the chemical, definite proportions and isomerism, now explained in terms of atomic structures in three dimensions.

It is strange that one of the crucial steps leading to this generally acceptable atomic theory was the discovery of the instability of some atoms; a discovery which led perforce to the abandonment of the notion of the indivisible atom. Henceforward, in defiance of etymology, 'atom' was used to mean chemical atom; and therefore many of the objections to atomism disappeared. In 1902 Rutherford and Soddy published a paper investigating the 'emanation' which was given off from thorium. In their 'General Theoretical Considerations', they remarked that radioactivity was generally thought to be an atomic phenomenon, and pointed to the significance of the fact that cathode rays are emitted in some radioactive processes, as β rays. They concluded that 'the radioactive elements must be undergoing spontaneous transformation';[90] and that the constancy of the rate of reaction, despite changes in external conditions, indicated that these changes were different in character from any that had been dealt with before by chemists: 'Radioactivity may therefore be considered as a manifestation of subatomic chemical change.'

In 1903 Soddy came to work with Ramsay in London and they began investigating the nature of the 'emanation', and the mechanism of subatomic chemical change. Radium bromide had recently become available commercially, and Ramsay bought twenty milligrams. The gas emitted was pumped away, and, after oxygen and hydrogen had been removed, its spectrum was examined. The characteristic helium line was present; but when the gas had been frozen in liquid air—which is not cold enough to condense helium—and evaporated again, no traces of helium could be detected. But four days later, when the spectrum was again examined, the helium line was visible. The 'emanation' had decayed and yielded helium; and this was the first transmutation to be followed in the laboratory. Ramsay and Soddy then went on and investigated the properties of the 'emanation', an inert gas to which the name 'radon' was later given.

Ramsay believed that radioactivity was a process akin to the spontaneous decomposition of an endothermic compound, that is, one in the formation of which from its elements, heat is absorbed. He therefore thought that the process might be reversed, and radium synthesised, by subjecting other elements to concentrated sources of energy such as radon. Thus in 1907 he said that: 'What we term the elements are, many of them, merely stable compounds which, before they begin to change, require to absorb a very large quantity of energy.'[91] The process of transmutation was probably taking place, he believed, in the stars. Despite all the developments of the previous fifty years, Ramsay's notions were not very different from those of Davy or Faraday.

In 1904 Ostwald, a friend of Ramsay's, gave the Faraday Lecture to the Chemical Society in London, and attempted to account for transmutation without introducing an atomic hypothesis.[92] According to Ostwald the various elements correspond to matter in regions of low potential energy. He proposed a picture of them as drops of water on stalactites hanging from the roof of a cave. In the region of the lighter elements the energy barriers are high, or the stalactites are long; so to transmute these elements, or to lift a drop of water to the top of one stalactite so that it may flow down another, requires much energy. But Ramsay's experiments with 'emanation' proved that in the region of the heaviest elements the stalactites were extremely short, mere corrugations in fact; and a drop on the 'emanation' stalactite could flow spontaneously over the low energy barrier and down towards the other end of the cave, where it would come to rest on the helium stalactite.

All this makes a pleasant picture to contemplate, but the objectives of this kind of positivistic chemistry must by 1904 have begun to seem excessively limited. Not only were its adherents denied the advantages of a theory which could explain the facts of stereochemistry, but they had also to forego the pleasures of atomic physics. For instead of accepting radioactive decay as a datum and merely designing a picture to help visualise it, physicists like J. J. Thomson and Rutherford were occupying themselves with the invention of atomic models in which the instability of the heavier elements, among other properties, would follow as a matter of course, and in which sub-atomic chemical changes could be explained. The plum pudding atom of J. J. Thomson was a step in this direction; and so was the later nuclear atom of Rutherford, who interpreted scattering experiments with α particles as indicating that the atom had a very heavy central nucleus, with electrons circling it in orbits. Bohr was able to extend this model to account quantitatively for the spectrum of hydrogen, and in a general way for other simple spectra; and also for the periodicity of the elements; and the Bohr atom proved capable of extension to give an account of valency. By the second decade of the twentieth century, scepticism towards atoms of the kind we have met in Brodie and Ostwald had become obsolete; and indeed Ostwald himself had been converted to atomism in 1907, the final push being given by Einstein's paper on the Brownian Movement. This effect, the random motion of pollen grains in water, was interpreted as due to the bombardment of the grains by the molecules of water in their ceaseless movement. This seemed to Ostwald as near as could be reasonably expected to an observation of molecules; and he therefore dropped his opposition to the concept.

An interesting half-way stage to the electronic theory of valency of G. N. Lewis, based on the Bohr atom, was Ramsay's theory announced to the Chemical Society in his Presidential Addresses of 1908 and 1909.[93] According to Ramsay, one should distinguish theories, which are candidates for truth, from hypotheses, to be judged by their utility, and fictions, which belong to the realm of art. The atomic theory seemed to him a real theory; and he proposed to extend it by an hypothesis, that electrons are atoms of the chemical substance electricity, possessing mass, combining with other atoms, existing in the free state as molecules, and serving as bonds of union between atoms. To see electricity on a list of chemical elements is surprising, and reminds us of Lavoisier's 'light' and 'caloric', and of the speculations of workers a century before Ramsay. But his notion was really quite modern. Since only negative electrons were known, he suggested that sodium, for example, was really a compound of sodion, Na^+, and an electron; its formula should be written NaE. Sodium is 'an orange of sodion' with 'a rind of electron'. When sodium combines with chlorine, the rind 'separates from the orange and forms a layer or cushion between the Na and the Cl, and ... on solution the electron attaches itself to the chlorine in some similar fashion, forming an ion of chlorine'. Ramsay went on to anticipate in a general way some notions of later valency theory, but that is another story.

In 1909 he began his lecture with the remarks from Faraday on the metals, concerning the deplorable increase in their number. His hypothesis this year was that the differences between the elements are due to their loss and gain of electrons; not the 'supplementary' ones which convert atoms to ions, but those at a deeper level: 'constituents of the atom, as it were'.

These remarks mark a stage in the abandonment of Lavoisier's definition of an element. Instead of describing an element in terms of existing analytical techniques, Ramsay would have spoken of the number of electrons in the core of its atom. When the nuclear or planetary atom had become accepted, Moseley studied the X-ray spectra of the elements and showed that they could be interpreted in such a way that one could derive from them the atomic number, or positive charge on the nucleus, of the various elements. And the modern definition of 'element' is in terms of atomic number, rather than of limits of analysis.

Ramsay pointed out that differences between elements and their ions are considerable. The loss of more electrons would therefore produce greater differences. Nor were such losses hypothetical, for electrons were emitted in radioactive decay, and Rutherford had followed radioactive changes through a whole series of substances having very different properties. Ramsay declared that

> 'where electrons are evolved, a change from one element to another is in progress',

and:

> 'Just as the concentration by Sir Humphry Davy of electrical energy into the small area of the ends of two platinum wires enabled him to decompose the alkalis, so a high concentration of energy will effect the change of one element into another, or into several others.'

The wheel had come full circle; Davy's conception of the chemical elements had become the accepted one, with the minute electrons in place of his Boscovich atoms. Nobody doubted now that matter was in some sense made up of atoms;

but the chemical atoms had become half-way stages in a hierarchy stretching from electrons (and very soon protons too) up to molecules of great complexity. Speculation as to the exact role of electrons in chemistry, and the mechanism of the evolution of the chemical elements, were the order of the day; but that electrons were in some way involved in valency, and that the elements were the products of an evolutionary mechanism, seemed beyond doubt.

Of the various traditions described earlier, the positivists were for the moment routed, though the philosopher/scientist Ernst Mach remained obdurate to the end in his agnosticism. It seemed possible to describe and explain the phenomena in terms of the unobservable atoms and electrons far better than without them. Atomic and molecular theories, and then theories of electrons and of atomic structure, could be made to generate testable consequences which were found to be in accordance with nature. These theories were so powerful as explanatory and heuristic devices that it was difficult to believe that they were completely fictional. After 1869 the positivist school in England seems to have become steadily weaker, although as long as Ostwald held out in Germany its views could not be altogether ignored.

Those who believed in the unity of matter were triumphant. Davy's prediction that definite proportions would be found to depend on the identity of the matters acting on one another had been verified in the electronic theories of matter and of valency. The discovery of isotopes proved that even Prout's conjectures had been correct; fractional atomic weights were accounted for when the elements were shown to consist of a mixture of isotopes all having approximately whole-number weights on the scale $H = 1$. Divergencies from exact whole numbers were explained, as Marignac had suggested, in terms of some other causes operating. Dalton's axiom that all atoms of an element weighed the same was dropped, and replaced by the proposition that they all have the same atomic number.

A mathematical chemistry was at last in sight, after the failure of Brodie's Calculus and the partial success of physical chemistry based on thermodynamics. As Davy had forecast, this kind of chemistry came about through the success of the adherents to the theory of the unity of matter. Subatomic physics, combined with thermodynamics, provided the basis for a mathematical theoretical chemistry; an edifice which is even now far from completeness, but is at least begun.

The theories of J. J. Thomson and of Rutherford have long been superseded but they form the basis of modern views in a way in which nineteenth-century theories do not. One of the main reasons for this would seem to be their acceptability to both physicists and chemists, or in other words, their extended range. The gulf which had existed between the disciplines of physics and chemistry had closed; and twentieth-century atomic theories have had to have a broad base in both sciences. One cannot help noticing the enormous divergences between this atomism and the views of the alleged reviver of atomic theories. For Dalton, the atoms of an element were all alike; whereas in modern chemistry most elements are composed of atoms of different weights, the isotopes. Dalton proposed simplicity rules for determining atomic formulae, and rejected the laws of gaseous combination; yet it was these laws which enabled agreed atomic weights to be arrived at in the 1860s. These weights led to the Periodic Table, and to clearer notions of valency, and then to stereochemistry, and made chemical atomism acceptable as it had not been in its first half-century, when definite proportions were all that it 'explained', and positivists were in the ascendant. One should not say that because a theory has changed in a century its founder deserves no credit; but when the divergences are as great as in this case, it seems that one can only say that twentieth-century theories are drawing on traditions other than the Daltonian.

Dalton again had been unclear as to whether his theory was a theory of matter at all, or simply a theory of chemical atoms; and this uncomfortable position was maintained by many of his successors. In so far as he had an opinion on the matter, it was that the elements were composed of indivisible atoms; another notion that had to go before the atomic theory became acceptable. The atomic theory had to account for physical as well as chemical phenomena before the more critical in any generation could bring themselves to accept it; otherwise they found themselves in the unpleasant position of using the theory and denying it at the same time, as Brodie put it. An author disposed to be polemical might well question whether the naïve atomism of Dalton had helped the cause of atomism in general; and might find himself having to admit only that, like Brodie's Calculus, Dalton's theory had aired the question, aroused controversy, and forced scientists to make up their minds when they might otherwise have kept them open. Dalton would, in this view, be a figure essentially of the second rank; a verdict in accordance with the general impression made by the *New System of Chemistry*, which, apart from the few pages on the atomic theory, is a work of very moderate interest. However, the attempt to classify scientists in ranks is probably not a profitable activity.

It is in the works of some of Dalton's critics that one finds not only a greater degree of anticipation of modern science— a fickle criterion of greatness—but also much more life. Davy and Faraday wrote books, papers, and lectures which can still be read with interest and pleasure by non-specialists, and raise all kinds of general questions. If there is to be a direct line of descent from theories of matter of the early nineteenth century down to the early twentieth and beyond, then it seems to be from Davy, recognised by his contemporaries as the most brilliant natural philosopher of his epoch, through Faraday, surely the greatest of his, and then through Crookes and others to J. J. Thomson. All these men looked for sub-atomic particles, or genuine atoms, of very simple properties, refusing to recognise the elements as ultimate, simple bodies. From Davy, whose researches with the new analytical tool, the voltaic battery, revealed a host of new elements, stems the theory that chemical bonding is electrical in character; a conclusion triumphantly vindicated in the electronic theory of valency. Crookes, in the discharge tube and the spectroscope, was also handling new analytical techniques; and his work pointed ultimately to J. J. Thomson's plum pudding atom, and on towards modern science. Sceptics such as Davy, Faraday, and William Thomson, asking, in the tradition of the corpuscular philosophy of Boyle and Newton, more of an atomic theory than Dalton's could provide, kept open room for dialogue between physicists and chemists; and ensured that chemists remained dissatisfied with a theory merely of chemical atoms, and physicists with the elastic particles of the kinetic theory.

This latter theory seems again to have derived little from Dalton; indeed, rather oddly, Davy, as an adherent to the kinetic theory of heat, seems to have been felt as a precursor by Graham. Dalton's theory of gases was certainly quite different from that of the kinetic theorists. Perhaps we should simply conclude that there were a number of questions which were answered at various times by theorists using various kinds of atomic theory; that these theories were usually incompatible with one another; and that it was not until into the twentieth century that a generally satisfactory atomic theory, with a wide explanatory range, became available. There is in the story no one reviver of atomism, for atomism did not need reviving in the nineteenth century; but Dalton's position is of some importance in that he convinced chemists of the reality of combination in definite and multiple proportions, and though his own was crude and naïve, aroused interest in atomic hypotheses as a method of explaining chemical phenomena.

References

1 I. Newton, *Opticks*, 4th ed., repr. New York, 1952, p. 400.
2 Ibid., p. 389.
3 Ibid., p. 394.
4 J. Locke, *An Essay Concerning Human Understanding*, esp. bk. II, ch. 8.
5 T. Garnett, *Outlines of a Course of Lectures on Chemistry*, London, 1801, p. 16.
6 *The Collected Works of Sir Humphry Davy*, 9 vols., London, 1839-40, VIII, p. 329.
7 R. Boyle, *The Sceptical Chymist*, Everyman ed., London, 1911, p. 223.
8 J. R. Partington, *A History of Chemistry*, vol. II, London, 1961, p. 665.
9 A. L. Lavoisier, *Elements of Chemistry*, trans. R. Kerr, Edinburgh, 1790, p. xxiv; facsimile reprint, New York, 1965.
10 R. Boscovich, *A Theory of Natural Philosophy*, trans. J. M. Child, Cambridge, Mass., 1966, §133.
11 R. Boscovich, op. cit., §99.
12 H. Davy, *Syllabus of a Course of Lectures at the Royal Institution*, London, 1802, p. 2.
13 J. Murray, *Elements of Chemistry*, 2 vols., Edinburgh, 1804; I, pp. 17, 31.
14 H. Davy, *Collected Works*, VII, pp. 93-7. See *The Journal of Gideon Mantell*, ed. E. C. Curwen, London, 1940, pp. 59-60.
15 Guyton de Morveau, *Nicholson's Journal*, I (1797-8), p. 110; G. Pearson, ibid., p. 355.
16 P. S. Laplace, *Exposition du Système du Monde*, 2nd edn., Paris, Au VII (1799-1800), p. 287.
17 *Alembic Club Reprints* II, 'Foundations of the Atomic Theory', p. 39.
18 W. H. Wollaston, *Philosophical Transactions* of the Royal Society, CIV (1814), 7.
19 H. Davy, *Phil. Trans.*, CVIII (1808), 1.
20 J. Davy, *Memoirs of the Life of Sir Humphry Davy*, 2 vols., London, 1836, I, p. 438.

REFERENCES

21 H. Davy, *Collected Works*, VIII, p. 323.
22 T. Beddoes, *Contributions to Medical and Physical Knowledge*, Bristol, 1799, p. 222.
23 H. Davy, *Collected Works*, VIII, p. 325.
24 H. E. Roscoe and A. Harden, *A New View of the Origin of Dalton's Atomic Theory*, London, 1896, pp. 100, 159, 49.
25 Ibid., p. 112.
26 W. Whewell, *A History of the Inductive Sciences*, 3 vols., London, 1837, III, p. 19.
27 A. Ure, *The Quarterly Journal of Science*, XX (1825), 113.
28 J. Herschel, *Preliminary Discourse on the Study of Natural Philosophy*, London, 1830, pp. 299, 305.
29 T. Garnett, *Outlines of a Course of Lectures on Natural and Experimental Philosophy*, London, 1801, p. 5.
30 C. Daubeny, *Introduction to the Atomic Theory*, Oxford, 1831, p. 21.
31 H. Davy, *Elements of Chemistry*, London, 1812, p. 489.
32 H. Davy, *Collected Works*, IX, pp. 383 ff., VIII, p. 271.
33 H. Bence Jones, *The Life and Letters of Faraday*, 2 vols., London, 1870, I, p. 217.
34 R. Chenevix, *Phil. Trans.*, XCIII (1803), 320.
35 H. B. Jones, *Faraday*, I, p. 256; this passage was quoted by Crookes in 1886 and by Ostwald in 1904—see Chapter 7.
36 W. H. Wollaston, *Phil. Trans.*, CXII (1822), 89–98. M. Faraday's paper was in the *Royal Institution Journal* for 1831. Both are discussed in T. Thomson, *A System of Chemistry of Inorganic Bodies*, 2 vols., 7th ed., London, 1831, I, pp. 4–6.
37 W. Whewell, *Philosophy of the Inductive Sciences*, 2nd ed., 2 vols., London, 1847, I, p. 436. He had put the argument to the British Association in 1839.
38 H. B. Jones, *Faraday*, II, p. 277.
39 M. Faraday, *Experimental Researches in Electricity*, 3 vols., London, 1839–55, II, pp. 284–93.
40 S. T. Coleridge, *The Friend*, footnote to pt. I, Essay 13.
41 M. Faraday, *Experimental Researches in Electricity*, III, pp. 447–52.
42 W. Whewell, *History of the Inductive Sciences*, III, p. 145.
43 W. Whewell, *Philosophy of the Inductive Sciences*, I, p. 387.
44 W. Whewell, *Philosophy of the Inductive Sciences*, I, p. 465.
45 W. Whewell, *Philosophy of the Inductive Sciences*, I, p. 432.
46 A. Koyré, *Newtonian Studies*, London, 1965, p. 35.

REFERENCES

47 C. Babbage, *The Ninth Bridgewater Treatise, a fragment*, 2nd ed., London, 1838, p. 32.
48 *Ibid.*, Appendix A; pp. 179–85.
49 R. Taylor (ed.), *Scientific Memoirs*, I (1837), p. 448 ff.
50 T. Exley's book was *Principles of Natural Philosophy*, London, 1829. His papers appeared in *The Philosophical Magazine*, third series, XI (1837), 496; and various *Reports of the British Association*; for 1836, p. 30; for 1838, p. 68; for 1844, p. 39; and for 1848, p. 50.
51 P. G. Tait, *Recent Advances in Physical Science*, 3rd ed., London, 1885, p. 295; Helmholtz's paper appeared in the *Phil. Mag.*, 4th series, XXXIII (1867), 485–512.
52 W. Thomson, *Notes of Lectures on Molecular Dynamics and the Wave Theory of Light*, Baltimore, 1884, p. 125; London, 1904, p. xi.
53 J. T. Merz, *A History of European Thought in the Nineteenth Century*, 4 vols., London, 1904–12, II, p. 57.
54 The vortex atom was proposed in the *Transactions of the Royal Society of Edinburgh*, XXV (1869), pp. 217–60. The remarks about chemists appear in S. P. Thompson, *The Life of William Thomson*, London, 1910, p. 517.
55 W. Thomson, *Report of the British Association*, 41st meeting (1871), lxxxiv.
56 J. J. Thomson, *A Treatise on the Motion of Vortex Rings*, London, 1883, § 59.
57. The kinetic theory is discussed in a series of papers by S. G. Brush, on which I have drawn heavily; they are in *Annals of Science*, XIII (1957), 188–98, and 273–82; and XIV (1958), 185–96, and 243–55. Herapath's paper appeared in *Annals of Philosophy*, New Series, I (1821), 274 ff, 340 ff, and 401 ff. I have quoted from p. 281. In 1847 he brought out his *Mathematical Physics*, 2 vols., London, 1847; the remark about chemistry having proved atomism is on p. 3. Maxwell's papers on the kinetic theory, and on atoms, may be found in *The Scientific Papers of James Clerk Maxwell*, ed. W. D. Niven, 2 vols., Cambridge, 1890.
58 T. Graham, *Phil. Mag.*, fourth series, XXVII (1864), 81.
59 R. A. Smith, *The Life and Works of Thomas Graham*, Glasgow, 1884, p. 107.
60 The Papers of Gay-Lussac and Avogadro, and Dalton's comments on the former, are reprinted in *Alembic Club Reprints*, IV, 'Foundations of the Molecular Theory'.
61 J. J. Berzelius, *Annals of Philosophy*, II (1813), 443 ff; III (1814), p. 51. Dalton's reply came in III (1814), 174; and Berzelius' rejoinder in V (1815), 122.

REFERENCES

62 B. C. Brodie, *Phil. Trans.*, CXL (1850), 804.
63 J. B. Dumas, *Traité de Chimie Apliquée aux Arts*, 8 vols., Paris, 1828, I, p. xxxv, and *Leçons sur la Philosophie Chimique*, Paris, 1837. For Dumas' anti-atomic phase, see G. Buchdahl, *British Journal for the Philosophy of Science*, X (1959), 120.
64 Translated in H. M. Leicester and H. S. Klickstein, *A Sourcebook in Chemistry*, New York, 1952, p. 325.
65 Ibid., pp. 264-5.
66 Ibid., p. 306.
67 J. F. W. Johnston, *Report of the British Association*, 7th meeting (1837), 173.
68 S. Cannizzaro, 'Sketch of a Course of Chemical Philosophy', *Alembic Club Reprints*, XVIII.
69 O. T. Benfey (ed.), *Classics in the Theory of Chemical Composition*, New York, 1963, p. 70.
70 Brodie's Calculus appeared in the *Phil. Trans.* CLVI (1866), 781-859; and CLXVII (1877), 35-116. The lecture 'Ideal Chemistry' appeared, with an account of the discussion after it, in *Chemical News* XV (1867), 295-305; and was reprinted as a pamphlet in 1880 (London). For more detailed references, see W. H. Brock and D. M. Knight, *Isis*, LVI (1965), 5-25.
71 W. Odling, *A Manual of Chemistry*, pt. 1, London, 1861, pp. 2 and 4; and *Outlines of Chemistry*, London, 1870, p. viii.
72 A. Kekulé, *The Laboratory*, I (1867), 303-4.
73 A. W. Williamson, *Journal of the Chemical Society*, XXII (1869), 328-65. For this address and the subsequent debate, see Brock and Knight, op. cit.
74 *J. Chem. Soc.*, XXII (1869), 433-41; *Chemical News*, XX (1869), 235-7.
75 J. Tyndall, *Fragments of Science*, 7th ed., 2 vols., London, 1889, II, p. 108.
76 E. Mills, *Phil. Mag.*, fourth series, XLII (1871), 112-29.
77 E. Divers, *Proc. Roy. Soc.*, LXXVIIIA (1907), xli.
78 A. W. Williamson, *Chemical News*, LVI (1887), 5-6.
79 R. Meldola, F.R.S., *Chemistry*, London & New York, n.d., p. 131.
80 H. E. Roscoe, *Report of the British Association*, 40th meeting (1870), 45.
81 G. Liveing, *Report of the British Association*, 52nd meeting (1882), 479 ff.
82 J. H. Gladstone, *Report of the British Association*, 53rd meeting (1883), p. 453.
83 W. Crookes, *Report of the British Association*, 56th meeting (1886), 558. See also *Alembic Club Reprints*, XX, 'Prout's Hypothesis'.
84 H. Helmholtz, *J. Chem. Soc.*, XXXIX (1881), 227.

REFERENCES

85 W. Crookes, *J. Chem. Soc., Trans.*, LV (1889), 256, 272.
86 J. J. Thomson, *Phil. Mag.*, fifth series, XLIV (1897), 302, 311.
87 In W. Tilden, *The Elements*, London & New York, 1910, pp. 81 and 97.
88 A. W. Rücker, *Report of the British Association*, 71st meeting (1901), 6–21.
89 W. Tilden, *The Elements*, p. 5.
90 E. Rutherford and F. Soddy, *Phil. Mag.*, sixth series, IV (1902), 395.
91 M. W. Travers, *Sir William Ramsay*, London, 1956, p. 257.
92 W. Ostwald, *J. Chem. Soc., Trans.*, LXXXV (1904), 506–22.
93 W. Ramsay, *J. Chem. Soc., Trans.*, XCIII (1908), 774–88; and XCV (1909), 624–37.

Bibliography

The bibliography which follows is not meant to be exhaustive. There is first a section of general works; and under the various chapter headings appear a few works from which quotations were not made, and which did not therefore appear in the references, but which are relevant to those chapters.

GENERAL

Much of what appears in this book has already been published in two articles, and those in search of a more exhaustive bibliography are referred there. They are: D. M. Knight, 'The Atomic Theory and the Elements', *Studies in Romanticism*, V (1966), 185–207, and W. H. Brock and D. M. Knight, 'The Atomic Debates', *Isis*, LVI (1965), 5–25. A compressed form of these two papers is to appear as the introductory chapter to W. H. Brock (ed.), *The Atomic Debates*, to be published by the University of Leicester Press towards the end of 1966.

A general history of theories of matter is S. Toulmin and J. Goodfield, *The Architecture of Matter*, London, 1962.

For *a priori* science, R. Harré, *The Anticipation of Nature*, London, 1965, is useful; and for a discussion of positivism, R. Harré, *Theories and Things*, London, 1961.

J. T. Merz, *A History of European Thought in the Nineteenth Century*, 4 vols., London, 1904–12; repr. New York, 1965, is very valuable; especially vol. I, chs. 4 and 5, and vol. II, ch. 6.

Extracts from original papers are reprinted in H. M. Leicester and H. S. Klickstein, *A Sourcebook in Chemistry*, New York, 1952; and in the invaluable *Alembic Club Reprints*; and in W. F. Magie, *A Sourcebook in Physics*, New York, 1935.

J. R. Partington's monumental *History of Chemistry*, vols. II–IV, London, 1961–4, may with advantage be consulted on occasion; as may M. P. Crosland, *Historical Studies in the Language of Chemistry*, London, 1962.

CHAPTER 1

A judicious selection from the writings of Robert Boyle, with an introduction, has recently been published: M. B. Hall, *Robert Boyle on Natural Philosophy*, Bloomington, 1965. Newton's views on atomism are explored in the relevant sections of I. B. Cohen, *Isaac Newton's Papers and Letters on Natural Philosophy*, Cambridge, 1958; and A. R. Hall and M. B. Hall, *Unpublished Scientific Papers of Isaac Newton*, Cambridge, 1962. Leibniz's views on atomism may be found in H. G. Alexander, *The Leibniz-Clarke Correspondence*, Manchester, 1956. For Boscovich, see L. L. Whyte (ed.), *Roger Joseph Boscovich*, London, 1961; and for Lavoisier, D. McKie, *Antoine Lavoisier*, London, 1952.

CHAPTER 2

J. Dalton, *A New System of Chemical Philosophy*, 3 vols., Manchester, 1808-27, has been reproduced in fascimile; London, n.d., 3 vols. in 2.
For a short biography of Dalton, see F. Greenaway, *John Dalton and the Atom*, London, 1966. See also A. W. Thackray, 'The Emergence of Dalton's Chemical Atomic Theory: 1801-8', *The British Journal for the History of Science*, III (1966), 1-23.

CHAPTER 3

On Wollaston's astronomical proof of atomism, see the article in G. Wilson, *Religio Chemici*, London & Cambridge, 1862. On action at a distance, Mary Hesse, *Forces and Fields*, London, 1961. And the very important new biography; L. Pearce Williams, *Michael Faraday*, London, 1965.

CHAPTER 4

On Babbage, see *Charles Babbage and his Calculating Engines*, ed. P. Morrison & E. Morrison, New York, 1961. An extract from his Bridgewater Treatise may be found in I. B. Cohen & H. M. Jones (ed.), *Science Before Darwin*, London, 1963. For a positivistic attack on atomism, see J. B. Stallo, *The Concepts and Theories of Modern Physics*, ed. P. W. Bridgeman, Cambridge, Mass.. 1960. (Reprint of 3rd ed. of 1888.)

CHAPTER 5

This period is covered in E. von Meyer, *A History of Chemistry*, trans. G. McGowan, 3rd ed., London, 1906; and in all standard histories. Some papers are reprinted in H. le Chatelier (ed.), *Molecules, Atomes, et Notations Chimiques*, Paris, 1922. See the article on the Carlsruhe Conference by Sir Harold Hartley, in *Notes and Records of the Royal Society*, XXI (1966), 56-63.

CHAPTER 6
See Brock and Knight, op. cit.; and W. H. Brock (ed.), *The Atomic Debates*, Leicester, in press.

CHAPTER 7
See M. W. Travers, *Sir William Ramsay*, London, 1956, for much interesting material on this period. Papers by J. J. Thomson, Rutherford, and others on electrons, radioactivity, and atomic structure are reproduced in facsimile in S. Wright (ed.), *Classical Scientific Papers, Physics*, London, 1964. Papers by Davy and Faraday appear in Cohen and Jones, *Science Before Darwin*, and enable the concern with general problems of these authors to be judged; see also the Introduction to that work.

Additional papers are reprinted in A. Romer (ed.), *The Discovery of Radioactivity and Transmutation*, New York, 1964; and D. M. Knight (ed.), *Classical Scientific Papers, Chemistry*, London, in press, will contain facsimiles of nineteenth century papers devoted to theories of matter.

Index

AMPÈRE, 87, 90, 92
Analogy, importance of, 41
Andrews, 124
astronomy, 20, 21, 31, 46–7, 59, 61, 63
atomic theory of Dalton, 2, 8, 14, 15, 16–36, 37, 46, 47, 48, 50, 53, 57, 59, 69
attraction and repulsion, 7–8, 16–17, 42, 54, 58, 62, 63, 64, 65, 66, 67, 68–9, 74, 78, 83, 90–1
Avogadro's law, 69, 82, 87, 88, 89–90, 91, 92, 94, 97, 99, 100, 103, 111, 116

BABBAGE, Charles, 61–5, 69, 70
Beddoes, Thomas, 28
Bel, le, 112
Berthollet, 22, 35, 83, 84
Berzelius, J. J., 31, 57, 86, 87, 91, 92, 95–6, 97
billiard ball theories, 6, 12, 13, 43, 92, 117
Black, 8
Bohr, 144, 145
Boole, George, 107
Boscovich, Roger, 9, 12–14, 38, 39, 40, 41, 58, 66, 67, 68, 70
Boscovich atom, 9, 12–14, 36, 38–43, 46, 50, 51, 52, 58, 62, 65, 66, 69, 70, 71, 74, 76, 79, 125, 138, 146
Boyle, R., 9, 10, 63, 138, 150
British Association, 44, 53, 66, 67, 72, 77, 95, 97, 98, 107, 112, 115, 128, 129, 135, 139
Brodie, Benjamin, 36, 92, 106–13, 114, 115, 119–20, 121, 122–3, 125, 133, 144, 149
Brown, Crum, 129
Bruno, 52

CALCULUS, Brodie's, 107–13, 115, 118, 119, 122, 125, 129, 148, 149
caloric theory of heat, 8
Cambridge University, 53
Cannizzaro, S., 25, 77, 89, 90, 93, 97, 103, 106, 109, 111
capillary rise, 55
Carlsruhe Conference, 77, 89, 97, 101, 103, 106
catalysis, 68
cathode rays, 45, 135, 137–9, 142
Cavendish, Henry, 65
chemical combination, 51–2, 53, 63, 116, 120
chemical molecular theories, 83–104
Chemical News, 107, 112

INDEX

Chemical Society, 34, 52, 104–26, 127, 136, 143, 145
Chemistry Meteorology and the Function of Digestion, 90
Chenevix, R., 45
Clausius, 76
Cohen, I. B., 4
Coleridge, S. T., 52, 122, 123
Collected Works (Davy) (quoted), 19, 28, 29, 39–40, 41–2
collision problem, 75
Comte, 129
conduction, heat, 49–50, 51, 55
Connexion of the Physical Sciences, 102
Consolations in Travel, 39–40, 43, 65
Contributions to Medical and Physical Knowledge (quoted), 28
corpuscular philosophy, 5, 6, 7, 9–10, 11, 12, 15, 16, 21, 33, 38, 42, 54, 137, 138, 139
Couper, 106
Crookes, William, 45, 46, 112, 132, 133, 134, 135, 136, 137, 138, 150
crystallography, 53–4, 57, 63, 97, 98
Cuvier, 40

DALTON, John, 2, 8, 14, 15, 16–36, 37, 46, 47, 48, 49, 53, 57, 68, 69, 78, 83, 84–7, 89, 90–1, 97, 99, 102, 106, 108, 117, 120, 121, 123, 139, 147, 148–51
Darwinism, 132, 134
Daubeny, Charles, 38–9
Davy, Humphry, 4, 8, 10, 14, 16, 18–19, 21, 24, 25–31, 33, 34, 35, 36, 37, 38, 39–45, 46, 48, 49, 58, 59, 62, 63, 65, 67, 74, 75–6, 81, 83, 87, 91, 95, 106, 113, 120, 123, 125, 137, 143, 146, 147, 148, 149, 150

Davy, Lady, 39
Davy, John, 27
de Rerum Natura, 5
Descartes, 125
Deville, 93
dimorphism, 97, 98
Donovan, 116
dualism, 91–2
Dulong, 101
Dumas, J. B., 90, 92–3, 94, 95, 106, 130, 131

EINSTEIN, 144
elasticity, 55, 63, 64, 65, 73, 76–7
electricity, 26–7, 29, 30, 42–3, 48, 49–50, 54, 62, 63, 65, 67, 135, 137, 139, 145, 146
elements, definitions of, 40–1, 133, 136, 146
Elements of Chemistry (quoted), 17, 25, 30, 39
electrochemical theories, 54
Eloge (of Don), 40
Epicurus, 5, 37, 130
equivalents, theory of, 49, 85, 86, 99, 101, 104, 116, 118, 122, 123
evaporation, 58
evolution, theory of, 132, 133–4, 137
Exley, Thomas, 65–70, 74
Experimental Researches in Electricity (quoted), 52–3

Familiar Letters on Chemistry, 100
Faraday, M., 4, 13, 14, 36, 37, 39, 43–54, 57, 58, 59, 62, 65, 66, 67, 80, 95, 96, 106, 120, 123, 125, 131, 133, 135, 143, 145, 149, 150
formulae, chemical, 105–14, 124

INDEX

Foster, Carey, 121–2
Frankland, 106, 113, 120
Frend, Dr., 54
Friend, The (quoted), 52

GARNETT, Thomas, 8, 16, 38, 39
Gassendi, 5
Gaudin, 87, 90
Gay-Lussac, Joseph Louis, 32, 35, 83–4, 85, 87, 88, 89, 90, 95
Gerhardt, 101, 105, 107, 108
Germany, chemistry in, 11
Gibbs, Willard, 140
Gladstone, J. H., 131, 132
Graham, Thomas, 80–1, 133, 150
gravitation, 7, 9, 13, 17, 20, 22, 53, 55, 61, 62, 64, 65, 66, 68, 74, 86, 121
Gray, Asa, 132
Grew, Nehemiah, 7
Guyton de Morveau, 21, 22

HARRÉ, Rom, 28
Hartley, Sir Harold, 111
heat, theories of, 8–9, 47, 63, 64, 81, 101–2, 113, 140, 150
Helmholtz, H., 70, 71, 135
Heraclitus, 52
Herapath, John, 74–5, 76
Herschel, Sir John, 35, 63, 106, 132
History of European Thought in the Nineteenth Century (quoted), 71
History of the Inductive Sciences, A (quoted), 34, 53–4
Hoff, van't, 112, 140, 141
Hooke, 24, 39

INSULATION, 51
Introduction to the Atomic Theory (quoted), 39
isomerism and isomorphism, 82, 95–102, 104, 112, 124, 127, 128, 142
isotopes, theory of, 136, 147, 148

JELLET, 129
Johnston, J. F. W., 98, 99, 100
Jones, H. B., 4
Journal of the Chemical Society, 139

KANE, Robert, 95
Kant, 125
Kekulé, atomic theory of, 2, 3, 32, 106, 113, 114, 115, 125
Kepler, 20, 31, 32, 120, 121
kinetic theories, 8, 9, 47, 60, 72, 74, 76, 77, 78, 79, 81, 111, 119, 129, 130, 131, 150
Kopp, Hermann, 102
Koyré, A., 53, 58

Laboratory, The, 106
Laplace, P. S., 22, 29, 55, 83
Laurent, 108
Lavoisier, A. L., 8, 11, 15, 16, 21, 22, 27, 29, 34, 38, 40, 41, 94, 108, 114–15, 119, 145, 146
legacy of the past, 5–15
Leibniz, 7, 125
Lewis, G. N., 145
Liebig, 94, 95, 96, 100, 106, 115
light, 44, 50, 64, 65, 72, 121
Liveing, G., 129–31
Locke, John, 8

INDEX

Lockyer, 133, 135
Lucretius, atomic theory of, 2, 5, 71

MACH, Ernst, 147
Mantell, Gideon, 18–19
Marignac, 130, 147
mathematical chemistry, 13, 20, 21–2, 30, 35, 36, 38, 39, 41, 55–6, 57, 58, 59, 61, 62, 63, 64, 65, 106, 138, 141, 148
matter, theories of, 11–12, 21, 29, 30, 33, 34, 43, 44–5, 48–51, 53, 60–82, 118, 119, 147, 148, 149
Maxwell, Clerk, 13, 38, 52, 56, 71, 76, 77–80, 111, 119, 132
Meldola, Professor, 128
Merz, J. T., 71
metals, 10, 45–6, 49–50, 145
metaphysical theories, 2, 4, 59
Miller, 120
Mills, Edmund, 121, 123–5
Mitscherlich, 96, 97, 98, 100, 101
monadism, 38
Moseley, 146
Mosotti, 60, 62, 64, 65, 66, 69, 102
multiple proportions, law of, 117, 120, 121
Murray, J., 17

Nature, 126
Naturphilosophie, 52, 123
New System of Chemistry, 32, 84, 149
Newton, Isaac, 5–7, 9, 17, 20, 31, 32, 33, 39, 44, 49, 54, 58, 61, 62, 66, 67, 68, 71, 81, 102, 121, 125, 138, 150
Ninth Bridgewater Treatise (quoted), 61

North British Review, 73, 112, 116, 120
nuclear theory, 138–9, 144

ODLING, William, 97, 107, 111, 113–14, 120–1, 126
optics, 61
Opticks, 5–7, 31, 49
Origin of the Species, 132
Ostwald, 36, 40, 126, 143, 144–5, 147
Outlines (Garnett) (quoted), 38
oxalates, analysis of, 23
Oxford University, 38, 107

PEARSON, 22
Periodic Table, 103, 128, 131, 133, 134, 136, 138, 148
Petit, 101
Phase Rule, 139
Philosophical Magazine, 39, 95, 123
Philosophical Transactions, 65, 102
Philosophy of the Inductive Sciences (quoted), 54–5, 57–8
phlogiston theory, 11, 108, 124
physics, 3
plum pudding atom, 139, 144, 150
positivists' theory, 20–1, 23–6, 32, 48, 56, 69, 73, 94, 106, 140, 147
Priestley, J., 39
Principia, 7
Proust, 83
Prout, William, 90, 91, 130, 131, 147

RADIATION, 44–5, 62, 142–3, 144, 146

INDEX

radical theory, 94–5
Ramsay, 79, 131, 136, 142, 143, 145, 146
Rankine, 102
Rayleigh, Lord, 79, 131
Roscoe, Henry, 112, 128–9
Royal Institution, 8, 14, 16, 26, 27, 38, 131
Royal Society, 18, 23, 75, 106, 107, 113
Rücker, Professor, 139–41
Rutherford, 142, 144, 146, 148

Sceptical Chymist, The, 10
Science Before Darwin, 4
scientific history, 1–4
Silbermann, 92
Smith, R. A., 80–1
Smith, Sydney, 59
Soddy, F., 142, 143
Somerville, Mary, 102–3
spectroscopy, 120, 131, 135, 136
Spencer, Herbert, 133
Stahl, 11, 54
Stallo, 73, 140
Stokes, 111, 112, 133
Stoney, Johnstone, 135
structural theory, 124, 126, 127
System of the World (quoted), 22

Tait, Peter Guthrie, 70, 71, 129
Tertullian, 125
theology, natural, 61
Theory of Natural Philosophy, A (quoted), 13, 14
thermal dissociation, 93
Thomson, J. J., 73, 74, 137–8, 139, 144, 148, 150

Thomson, Thomas, 19, 20, 26, 35, 87
Thomson, William (later Lord Kelvin), 70, 71–3, 76, 78, 79, 112, 118, 129, 141, 150
Tilden, Sir William, 139, 141
transmutation, 143
Treatise on the Motion of Vortex Rings, A, quoted, 73
Tyndall, John, 121

Ure, Andrew, 35

Valency, 9, 73–4, 97, 103, 106, 112, 113, 117, 135, 144, 145, 147, 148, 150
vaporisation, 46, 47, 92–4
vortex atom, 60, 61, 70–4, 78–9, 118, 130, 138, 139, 141

Wald, F., 139
Waterston, 76, 78, 113
weights, atomic, 18, 19, 22, 23, 24, 25, 34, 35, 41, 49, 86, 93, 97, 98, 99, 100, 101, 102, 103, 106, 113, 116, 124, 130, 131, 132, 134, 135, 136, 147, 148
Whewell, William, 11, 13, 34, 36, 38, 47, 52, 53–9, 61, 72
Williams, Pearce, 43, 46
Williamson, A. W., 3, 34, 105, 106, 113, 115–21, 125, 126, 127–8, 141
Wilson, George, 47
Wöhler, 95, 96
Wollaston, W. H., 19, 20, 21, 23–4, 25, 26, 27, 32, 35, 42, 45, 46–7, 49, 53, 58, 69, 106, 122, 125